Casement
of Juniata

—*From the Saddle and Sirloin Club Portrait by Joseph Allworthy.*

Capt. Dan D. Casement

Stockman, Farmer, Patriot, Exponent of the
Pioneer Self-Reliance, Personal Freedom and
Individual Independence
That Make the American Tradition.

Casement of Juniata

As a Man
and as a Stockman . . .
One of a Kind

DONALD R. ORNDUFF

With an Introduction by J. Evetts Haley

The Lowell Press/Kansas City, Missouri

©Copyright 1975 by Donald R. Ornduff
All Rights Reserved
Printed in the United States of America
Library of Congress Catalog Card No. 75-29800
ISBN 0-913504-27-0
First Edition

To Friendship

"...than which there is no finer wine"

Foreword

It bespeaks something of the magnetism of a man and the imprint he leaves when his name continues to evoke warm response long after he is gone. Such is the case with Dan D. Casement.

Widely hailed as one of the best loved stockmen of his time, his friends during a long and fruitful lifetime were legion. They appreciated greatly his obvious talents in the production of livestock of several breeds, his skills which were demonstrated by the victories won by his entries in carload cattle competition over a period of some 45 years, and his genuine love for the soil which nurtured the crops and grasslands that enabled his animals to thrive.

Important as these elements were, they were only the more obvious parts of the make-up of this extraordinary man who lived a full and adventuresome life and who personified to a marked degree the courage, romance, simplicity and integrity of earlier times in pioneer America.

Many who are yet active remember Dan Casement with affection. For them, and for the enlightenment of a newer generation which was denied the privilege of his acquaintance, it is hoped here to present additionally some insight

CASEMENT OF JUNIATA

into the nature, character and personality of a man who long ago became a legendary figure in the American livestock industry.

It was because he was the kind of a man he was that Dean Krakel, managing director of the National Cowboy Hall of Fame and Western Heritage Center at Oklahoma City, suggested that an article on Dan Casement be written for one of the 1974 issues of *Persimmon Hill*, the official publication of that prestigious organization. A subsequent article probing other aspects of the Casement career was published later in 1974 in the *American Hereford Journal*, a magazine devoted to the breed of beef cattle of which Casement was a life-long adherent.

This book mainly comprises the substance of those two articles, plus some other material and several illustrations. Appreciation is extended to these publications for the privilege of this presentation. Co-operation which likewise is appreciated was given by the Kansas State Historical Society, the Special Collections Library of Kansas State University, Orville B. Burtis, Sr., of Manhattan, Dr. A. D. Weber, also of Manhattan, and Mary Casement Furlong, of Painesville, Ohio.

—D.R.O.

Reverie and Reflection

By J. Evetts Haley

If Don Ornduff, in his long and able editorship of the *American Hereford Journal,* had done no more than induce Dan Casement to write his recollections, he would still have done enough to merit the lasting gratitude of all lovers of the literature of the Western Range.

But Mr. Ornduff, after retirement from the long and gruelling years of filling the pages and meeting the deadlines of that massive *Journal* dedicated to promoting the beefy, prepotent virtues of the royal, white-faced bloodlines, turned his mature talent and his energies toward history. His industrious scholarship has given the reading public such fine books as *The Hereford in America*, his historical overview in the singularly beautiful volume, *Bell Ranch as I Knew It,* and now this warmly appreciative study of that fiery and uncompromising American, formidable adversary, ardent lover of horses, tolerant friend, and compassionate companion of cows—the unforgettable Dan Casement of Kansas.

Shortly after his publication of the Casement memoirs in serial form in the *Journal*, the first segment under the heading of "Random Recol-

CASEMENT OF JUNIATA

lections," and a subsequent series of "Juniata Jottings," Mr. Ornduff reprinted the recollections in a modest little book long since out of print and now avidly sought by cow country collectors. These essays, along with Mr. Casement's earlier and excessively rare *Abbreviated Autobiography of a Joyous Pagan*, are not only of exceptional literary merit but furnish some of the finest insights into life on the range, its characters and its healthy, wholesome philosophy, of anything that has been written.

Somehow, through the years, the lovers and critics of this field of literature seem more or less to have overlooked the importance of Dan Casement's recollections. But in my opinion, for whatever it may be worth, I unhesitatingly rank them among the best in western writing and cowboy literature. And if I am prejudiced—a healthy human trait I cultivate and dearly cherish in this case—it is because of my deep affection for and admiration of the man.

Likewise I shared with him the oft-remembered experience of what terrific toil it is to translate thought and feelings into that lasting testament, and what should be almost sacred in form —that ineradicable and irrefutable evidence against the writer of his own words in print. In the popular sense, inspiration in writing at times undoubtedly has a part, though I suspect that good writing itself is more likely to spring from a deep and moving interest in the subject.

That fine, genial and great chronicler of the range, Andy Adams, once observed as we casual-

An Introduction

ly discussed some such point, that at times "a certain fine frenzy" seizes upon a writer, which, "as long as it lasts," is highly conducive to literary production. But through the years I have found that writing worthy of the printed page is mainly a matter of rigorous discipline and hard work.

Those moving words on a page of print, like the sensitive line and tone of color on a master artist's canvas, were not just set down with pen and ink, but were first ground out in sweat and blood, which in the end may have infused that work of art with inspiration, and made it true and lasting.

Dan Casement's essays, most of which were written when he was a seasoned, sensitive man of many years, *were* inspired by a cultivated background, a long life of rich experience, fine lust for adventurous action, courageous moral purpose, worthy and distinguished accomplishment, and the precious memories of men and times that were gone. Yet he lustily swore that they were ground out in hard work.

Mainly from these verdant memories, stimulated into fluent expression by Don Ornduff's persistent prodding, came the raw material for this book to limn Dan Casement's vivid figure upon the horizons of our Western history. Both Don and I knew and loved him dearly.

Good Lord! What went with all that time— never laggard when eager horses are our mounts. It has now been more than forty years since Dan Casement hunted me up at a cattlemen's as-

CASEMENT OF JUNIATA

sembly in Fort Worth to express his hearty approval of the substance, philosophy and style of an article I had written for a spring issue of *The Cattleman,* that fine publication of the Texas and Southwestern Cattle Raisers Association. That meeting began a friendship that grew warmer with every season until—after more than eighty joyous and adventurous years of life for him—the dread malignancies of the flesh robbed this tense world of the benign presence of his brave and irrepressible spirit.

In the crowded, tempestuous time since that meeting in the lobby of the Blackstone Hotel—a once cherished rendezvous of old-time cowmen that passed with them and Fort Worth's vaunted boast of being their town—most of my own time and interests have been spent among and devoted to the work of the sunburnt men and lands of grass. And almost always, looming strong among those silhouetted memories of significant events and colorful characters has been the figure of that cowman from Kansas, Dan Casement of the lovely Juniata.

As the years lope by and nature quirts and spurs me along with the inescapable disciplines of her moods and seasons, in the way of a stiff and stove-up old cowman I prowl our droughty ranges at a slow pace and now always on a gentle horse. Sometimes in that blessed solitude, I throw off the dreary realities that rise in waves of heat from the sun-blasted soil, and find myself riding dreamily with those colorful characters whom I have known in the distant past.

An Introduction

And the shades of many great men rise up to ride with me, especially that of . . .

Dan Casement! As long as I live that name and the memories which his rare and colorful character provoke will come back to enthrall and inspire me. Cowboy, horseman, livestock breeder, master feeder, embattled patriot and generous companion—in every facet and aspect of his versatile and accomplished nature, he was an extraordinary man.

How fortunate for those of us who cherish the undiminished color and glory of the ranges of grass that Don Ornduff has given us this worthy little book on this truly great and genuine man who for so long mingled and counseled with ranch people across western America and always rode straight in the saddle among his beloved Flint Hills of Kansas.

JH Ranch
Canyon, Texas

Contents

Foreword	VII
Reverie and Reflection (An introduction)	IX
Part I	1
A Casement Family Album	33-56
Part II	57

Part I

If it is trite to say that Dan Casement became a legend in his own time, so be it! He did. And in the field of agriculture and livestock production where extraordinary achievement seldom is broadly acknowledged. He earned recognition far beyond the limits of the business in which his extraordinary talents and boundless energy were concentrated for most of his long and productive lifetime. This he did through the force of his personality, the breadth of his interests, the vigor with which he expounded his always firm convictions, the colorful flow of his conversation and, of course, his actual accomplishments on the land.

He was a man's man, yet when the occasion was appropriate he could be the most cultivated of gentlemen. His manner was an attractive blend of courtliness and friendliness—characteristics which for decades made his Juniata Farm, near Manhattan in east-central Kansas, a Mecca not only for stockmen but also for a broad range of others drawn there by the warmth and magnetic personality of this unique man.

Perhaps no better description of Dan Casement exists than was devised in 1937 by Dr. F. D. Farrell, his long-time neighbor as president of Kansas State Agricultural College, the education-

CASEMENT OF JUNIATA

al institution at Manhattan now known as Kansas State University. Dr. Farrell, in a prepared introduction of him, said:

"He is an aristocrat in the best sense. He is also a democrat, spelled with a small—a very small—'d.' He is equally at home and happy in a farrowing pen and at the Waldorf-Astoria. He is intensely public spirited and equally intensely individualistic. In many respects he is a liberal, but in economics and politics he is a fundamentalist. He believes devoutly in the survival of the fittest but is among the first to lend a helping hand to human derelicts who are the embodiment of unfitness.

"His written English has almost Addisonian elegance and purity. His spoken English varies slightly with his mood and with the occasion. Sometimes he talks like an angel. At other times he talks differently. He knows and loves the songs of birds and the highly spiritual poetry of Allen Seeger, but, if pressed sufficiently, he lifts profanity almost into the realm of the fine arts.

"In short," Dr. Farrell concluded, "he is scholar, farmer, stockman, poet, sportsman, philosopher, statesman, individualist, altruist, patriot. He is Dan Casement, and there is only one of him."

J. Evetts Haley of Texas, eminent historian of the Southwestern range country, who has known most of the leading cowmen of the past half-century, said in 1974 that among the many hundreds of ranch people he has met during that

As a Man and as a Stockman

span there had not been another like him. As Haley put it, "After Dan Casement was cast, the mold was broken."

In 1949 Haley spent a memorable day with the Kansan, whom he described in a subsequent article, "A Day With Dan Casement," as "a notable man from any angle, who shines with a rough and ready brilliance from every facet of his many-sided nature. It is now some 20 years," Haley's article continued, "since this individualistic man with his booming voice, the ready wit, the hearty laugh, the deep sympathy and compassion and the picturesque profanity, hunted up a strange youth in the crowded cowtown of Fort Worth, during a Fat Stock Show, to compliment and wish him well. Never had my struggling efforts at authorship been blessed with such a panegyric of profanity. Secretly, I glowed for years beneath his unique and really reverent benediction."

How appropriate, upon reflection, seem Haley's revealingly descriptive words as applied to Casement upon the occasion of that 1949 visit:

"This charming soul of over 80 fruitful years who flourishes and fulminates, who battles well and bravely lives on hard but loving labor," Haley wrote, and then went on, "Nothing about this man is mean or ordinary." At another point he referred to the "choppy, booted walk" of Dan Casement, "impassioned and embattled philosopher—without any contradiction in terms —of the lovely Juniata." Yet another reference

CASEMENT OF JUNIATA

to his visit in Kansas with the man whose eyes even yet radiated a "bright blue glow," cited Casement as a "lusty combination of high idealism and practical cowman, cultured living and frontier ruggedness, violent adversary and gentle, generous friend . . . What an unbelievable combination of intelligence and temperament, sense and sentiment, of character and courage, still riding whip and spur upon his rugged frame! Truly, blood will tell!"

Dan Casement's family inheritance—blood line may be more exact—is as fascinating as was the man himself. His grandfather, Robert Casement, came to America as an impoverished immigrant late in 1828 from the Isle of Man, a small dot of land in the Irish Sea, about equidistant from Ireland and England, and slightly south of the southern line of Scotland. Said to have been an instinctive builder, with ancestors who had constructed a great water wheel at Loxey and the mercantile piers at Southampton, England, he located first at Geneva, New York. There, within three months of the family's arrival, was born a son, John Stephen Casement, who, in due course, fathered the son who was christened Dan Dillon Casement. Through his father's family Dan Casement thus was technically a Manxman, but somewhere back along the line of his Manx ancestors he inherited the Irish temperament and their wit, blarney and loveableness. Scotland also seems to have con-

As a Man and as a Stockman

tributed to his inheritance, and from the Scots is said to have come his love of animals, his thrift and an instinctive canniness that could have come from nowhere else. England is said to have been the source of his inherited gentility, sense of good sportsmanship and the ability to face and surmount hardship.

Casement's maternal grandfather was Charles Jennings who had migrated in 1820 as a boy from Vermont to Ohio. In 1840 he bought the farm afterwards known as Jennings Place where the daughter, Frances, who was to become Dan Casement's mother, was born in a log cabin that same year. It was in the "old house" on the farm, near Painesville, just east of Cleveland, Ohio, that Dan Casement was born on July 13, 1868. Three years later the family moved into a handsome and stately new home on this same place.

Grandfather Jennings, Dan once said, was a forceful personality—a man who "hated evil fervently and was ready to fight for righteousness at the drop of the hat." In writing of his mother many years later, the son observed that her "character reflected the strength of the granite hills of Vermont, with the soil of which her forebears had battled for generations, but was tempered by great tolerance, sympathy and human understanding." It may be significant, too, that her close friends included Susan B. Anthony and Elizabeth Cady Stanton, leading feminists of their time.

The 10 children in Robert Casement's family

CASEMENT OF JUNIATA

were reared in what was later described as abject poverty, but the son, John Stephen, did not let the fact that he was underprivileged and that his formal schooling was restricted to a basic knowledge of the three R's deter his ambitions. By the age of 10 he was working as a water boy with railroad construction gangs, and by the time the family moved to Michigan from New York while he was in his mid-'teens he was holding down a man's job strapping down iron rails on wooden stringers. When the family was left fatherless there he used the income from his labor to put a roof over his mother's head before he would consider buying an overcoat for himself. By the age of 29 he had applied himself so earnestly as a common laborer that he was made foreman of a track-laying gang on an Ohio line which later became a part of the New York Central Railroad.

When the Civil War came, John Stephen Casement joined the 7th Ohio Volunteer Infantry and threw himself into the conflict with such verve that he shortly became a colonel and later was breveted brigadier general for distinguished conduct in battle. For his feats during and after the war, he came to be widely hailed as General Jack.

His post-war accomplishments centered upon a project in which he was joined by his brother Daniel—that of doing the greater part of the grading for the road-bed and laying all of the track for the Union Pacific Railroad between Omaha and Utah. The Casement brothers in-

As a Man and as a Stockman

geniously devised something never before seen —a railroad work train—which contributed remarkably to the ability of their crews in putting down as much as six to eight miles of track daily in pushing the line through to the historic meeting with the Central Pacific at Promontory Point, Utah, in May, 1869.

Another first for the Casement brothers, and one that was highly popular with the hundreds of men working for them, was their scheme for providing fresh beef as a staple in the meals for their crews. A herd of cattle, mostly Durhams and Galloways, and branded with a "C" enclosed in a diamond, was assembled and ranged along the iron trail as it snaked its way westward, providing fresh beef all the way. Not only did this help in keeping their crews fully staffed but it also proved a point. By enduring three severe winters on the plains without heavy losses, the Casement herd provided evidence that cattle could be handled there under outdoor range conditions during even the most severe weather. Previously the westward emigrants had trailed their cattle through the region only in the summer months. The Casement example was not lost upon those who subsequently turned the country which the railroad traversed into a vast cattle domain.

It is thus clearly evident that it was from uncommon stock that Dan Casement sprang. As Dr. Farrell expressed it, "The ability and versatility that Dan possesses are not transmitted by human scrubs . . . but rather bespeak an ancestry of strong and distinctive character." This exemplified a basic tenet of Casement philosophy in the field of livestock production —that sound ancestry is vital to successful accomplishment. As he once put it, "Blood lines mark the foundation pattern of the individual, whether it be for the human animal or for his brute brother out in the pasture. Breeding is only half the battle, of course," he went on, "but it is always the first half."

Having been denied the advantages of much in the way of formal schooling, General Jack Casement saw to it that his son Dan, in due course, was properly enrolled at Princeton College (later University), from which he was graduated in 1890. While there he demonstrated that he was a man of action by playing both baseball and football, serving as captain of the second football team and as president of the baseball association. Though of only small stature, he pulled tug-of-war on the varsity team. He was on the editorial board of the university newspaper and served as class president during his junior year.

Besides, he exercised to the utmost his unusual talent for making friends. In fact, many years later he was to write that "Friendships were the most valuable acquirement of my college years." He went on to study law and political science at Columbia University in New York City which granted him a Master's degree in 1891. This course, he afterwards commented,

As a Man and as a Stockman

"was dictated by no insatiable thirst for knowledge but rather by a nostalgic yearning to prolong my boyhood."

It was with one of the great friends of his boyhood, Charles A. (Tot) Otis, Jr., of Cleveland, who had graduated from Yale University before the two became roommates while at Columbia University, that the young Dan Casement first engaged actively in ranching. In the early 1880's the fathers of the two had come into possession of a tract of range land in far-western Colorado. It was in the remote Unaweep Cañon, some 40 miles below the town of Grand Junction, and around 15 miles east of the eastern boundary of Utah. Dan went with his father to see it for the first time when he was a lad of 15 —a land from which the Uncompahgre Utes had just been moved into Utah—and he never forgot it. "I succumbed completely," he was later to say, "to the charm of a virgin country untouched by civilization."

Casement, who was 22 years old when he went there to commence his ranching career, once described the region in these words:

"The Unaweep Cañon is absolutely unique in that it cuts a deep gash through the Uncompahgre Plateau for a distance of 50 miles from the Gunnison River to the Dolores and has a distinct divide about midway of its extent. At this point the cañon is nearly a mile wide and quite a half-mile deep.

"Rising here, East Creek flows into the Gun-

CASEMENT OF JUNIATA

nison, and West Creek into the Dolores. The cañon takes its name from a Ute word signifying 'dividing of waters.' The walls of the cañon at both ends are of red sand rock but in most of its extent it cuts deep into the underlying structure of grey granite, so, for many miles on either side of the divide, the lower half of the walls are of this material. Above the granite a continuous bench or offset marks the base of the sand rock cap and testifies to its greater susceptibility to erosion.

"The northern wall of the cañon, more directly exposed to the sun, is rough and sparsely timbered with juniper and piñon, types indigenous to lower altitudes, while the opposite wall, where the snow lies deep in winter, is less rugged and supports a denser growth of oak brush, aspen, pine and spruce—typical growths of that region at altitudes between seven and nine thousand feet, the actual elevations encompassed by the wall from the base to summit.

"Three miles west of the divide, at the ranch site, Fall Creek, in spring, when the winter's snow is melting on the high mesa above, tumbles over the granite in an 800-foot cascade. The cañon's whole extent affords unusual scenery of surpassing beauty.

"Probably such a minute description of the Unaweep [which comprised some 3,000 acres on the floor of the cañon, and with which property proper went rights to use in common with others a range of over 100 square miles] can be explained and justified only by the fact that

As a Man and as a Stockman

nearly sixty years of my life are closely linked with this locality. Here I spent my most formative years and obtained the most important part of my education in the business of living. Here I learned the dignity and delight of hard, manual labor; the joy of life supported only by the most elemental essentials; the deep satisfaction that accrues from finishing tasks involving hazard and hardship.

"Complete dependence on the horse for every economic and social function incident to living gave me an invaluable appreciation of equine nature and not infrequent nobility. Here I gained something of tolerance and understanding by sharing fully the lives of natural men who lived simply and, for the most part, bravely, who had gained their practical wisdom by experience, who saw life objectively, whose actions and convictions were ordered by sound, age-old institutions and by reason which ascribed a purely biologic basis to the truths by which successful and satisfactory lives are invariably lived."

There, on the Unaweep Cattle Range, as the two city boys just out of college named it, their Triangle Bar outfit experienced all of the vicissitudes of pioneer ranching, including depredations of rustlers, sparse forage, low prices and the like. In a land largely empty of human habitation and entirely empty of any semblance of law, they learned to carry pistols and how to use them. But the two young men found the life highly stimulating, even though such convivial Grand Junction institutions as the Windsor, the

CASEMENT OF JUNIATA

Senate, the Board of Trade and the Bucket of Blood—the social and business centers of the town and surrounding region—were a bit too far distant to suit their fancy. But they didn't consider this too much of a hardship.

The rustling which had plagued their operations from the outset continued after Otis relinquished his interest, and the losses mounted. Finally, when 30 cows were driven off on a winter night, following a long series of cattle thefts, Casement became so incensed that later accounts credited him with catching the rustlers almost single handedly and seeing to it that three of them were put behind bars. Thinking back to the time when he faced up to the "hard and hateful" task of testifying against them in a community which seemed to be more or less indifferent to legal processes, resulting in what was said to have been the first conviction for cattle stealing ever recorded in Colorado's Mesa County, Casement later said that nothing up to that point ever brought him "fuller consciousness of a positive duty done."

Normal product of the Unaweep breeding herd operations consisted of three-year-old steers, the first shipment of which to the Omaha market, Casement once recalled, grossed less than three cents a pound. But it was not the money so much as the experience gained that counted, Casement said in reflection many years afterward. It was in western Colorado, he said, that he learned something of cattle and horses, "but

As a Man and as a Stockman

what was far more valuable, I learned to know the men who herded the cattle and rode the horses. I had the good fortune to see the last of the old days and to share the lives of great men who lived simply and naturally and made a priceless contribution to the best tradition of America." But this did not close his mind to the potential of the "new days" as he moved forward in the cattle industry of the 20th century.

After taking over the Otis interest, Casement also started a modest-sized registered Hereford herd to produce bulls for sale to other ranchers. This was maintained by Dan Casement until 1931 when the Unaweep was deeded to his son Jack upon the latter's marriage. Jack Casement ranched there for 10 years before selling the ranch and moving his Hereford cattle and Quarter Horse operations to eastern Colorado, where he died in 1972, leaving two daughters but no son to carry on the family name.

In retrospect, Casement once wrote, in thinking of the "old days," it seemed that activities centering on the ranch could have left no time for other interests. "The fact is," he went on, "that each fall, discarding chaps and spurs, Tot and I hurried east in order to attend the Yale-Princeton football game, to refresh old friendships and briefly to pound Fifth Avenue in frock coats and silk hats." It was on one of these sojourns that the two young dandies attended a Wild West Show. The story goes that when one of the riders was unceremoniously dumped by a plunging horse the two rather loudly chided him. The cowboy, thinking to quiet them, walked as close to their seats as possible and invited either one of them to do better if he could. Casement, by then no tyro when astride a horse, immediately accepted the challenge. He handed his top hat to his partner, climbed aboard the excited horse and, with the claw-hammer tails of his coat flapping with each jump, proceeded to ride the mount to a standstill. To the loud applause of the crowd.

Whether or not this tale is entirely factual, anyone who ever knew Dan Casement will appreciate that it could be.

This pattern of life ended for Casement with his marriage in December, 1897, to Mary Olivia Thornburgh, whose father, Major T. T. Thornburgh, commandant at Fort Fred Steele in Wyoming, had been killed when she was five years old in a battle with the White River Utes in Colorado. The newlyweds went immediately to Costa Rica where the General had contracted to build a railroad in that Central American country. Young Casement had been called upon to help in what he once described as "a desperate venture," undertaken because the General was "out of money." Although he had earned substantial sums in other railroad construction projects which followed upon his spectacular Union Pacific success, he also had an apparently well-deserved reputation for dispensing his resources freely, if not extravagantly. So much so, in fact, that the eulogist at his funeral remarked

As a Man and as a Stockman

that he seemed "to delight in gathering money in order that he might scatter it with both hands." In the light of that statement, it is of interest to recall that he came into possession in 1878 of the section of bottom land near Manhattan which was to become the nucleus of Juniata Farm in settlement of a debt, and that his involvement with the Unaweep range came about because he lent financial backing to a land-promotion project.

The Costa Rica venture, on which the General, his wife, and their son Dan staked everything they possessed, was a financial success, and the profits, at the end of the six years involved, were divided evenly between the General and his son. This sense of prosperity led the younger Casement family, by then including a baby daughter, to spend three years rather idly in the East and South, quail and snipe shooting in Georgia, fishing in Virginia, playing polo, wintering in Washington, and, in general, doing whatever caught his fancy, or, as Dan put it, behaving "more or less like a playboy."

The family next took up residence in Colorado Springs where he played polo and moved easily in the social circles for which that Rocky Mountain city in the shadow of Pikes Peak was noted. Casement rationalized this decision on the basis that "the Springs," as the town was widely and familiarly known, was "about half way between the Kansas farm and the Colorado ranch," and that thus located he could more readily visit each frequently.

CASEMENT OF JUNIATA

After residing in Colorado Springs for nine years, with his family by then including a son and a second daughter, Casement faced up to the fact, as he once expressed it, "that in this fashion of life outgo was exceeding income. Besides," he went on, "I realized that, in order to preserve character and self-respect, I would have to attend more closely to my business," the potentially most valuable asset of which was Juniata Farm—so named for the old Juniata crossing of the Big Blue River there. Thus, in the fall of 1915, the Casements loaded themselves in the family's Stevens-Duryea motor car and headed for Manhattan. A small house in town was rented, the children put in the public school and, as the head of the family later said, "Thenceforth our lives took on a larger measure of responsibility and more definite direction," adding reflectively, "From that time I knuckled down to the job and have been in the business head over heels ever since."

Evidently there had always been deep within him an instinctive affinity for livestock and the land. This had led him as a lad, when the subject of his future occupation arose in a conversation with his father, to express a desire "to farm," in consequence of which the General deeded Juniata Farm to him as of his 21st birthday. The property included not only the original section of crop and meadow land along the river but also around 2,400 acres of hill pastures—the Bluestem Hills, across the river and well up from it—which the General had acquired by 1880.

As a Man and as a Stockman

But just as Casement was getting well under way at Juniata with his Herefords, Ayrshires, Suffolk-Punch and Quarter Horses, two or three breeds of hogs and a flock of sheep, World War I intervened. Having regretfully missed the Spanish-American War because of his obligations in Costa Rica, as he later expressed personally to President Theodore Roosevelt, he volunteered for active duty in 1917 despite the fact that he was then 49 years old. "I felt that it was my duty to get in," he later said. He was commissioned a captain of field artillery and once recounted something of the discipline involved in learning to take orders and to say "sir" to a major 25 years his junior. He also recalled that "it was torment" to stand at attention for long periods of time, but, he continued with what seemed to be a sense of pride, he "did it."

At length, after completion of training at Fort Sheridan, he and his company of replacement troops embarked for Europe on the *Tuscania*, and he barely escaped with his life when the ship was torpedoed off the northeast coast of Ireland. He was in the midst of the rescue effort and was credited with having demonstrated unusual valor in the lifesaving effort. "After we had launched all of the lifeboats," he once related, "those of us for whom no boats or rafts were available awaited submersion with such fortitude as we could severally summon." In the nick of time, the little British destroyer, *Pigeon*, sneaked in out of the darkness and took off those who had remained on the *Tuscania*, with Case-

CASEMENT OF JUNIATA

ment being the next to the last man off the transport, it was said. When the scattered survivors were reassembled at Londonderry, it was found that more than 200 lives had been lost.

The same brand of courage was evidenced many years later when a hold-up man confronted Casement at the Stock Yard Inn where he was staying while attending the Chicago stock show in December, 1950. Despite the fact that he was then well past 82 years old, he refused to submit and created such an uproar that the would-be bandit fled in confusion. It is virtually certain that he found zest in his war-time experiences and his command of the second battalion of the 27th Field Artillery in France, considering them more or less, and perhaps unconsciously, as an extension of the adventures of his earlier years. And he expressed regret that his unit did not get into the thick of the fight.

Although his first 50 years had been a joyous swirl, as he once expressed it, he now was ready to settle down. And there can be no doubt that he found his later years as a nationally recognized farmer-stockman fully as pleasurable and satisfying as his earlier life had been.

Although he had exhibited carloads of fat steers at the American Royal Livestock Show in Kansas City as far back as 1907, and had won the championship award there in 1908 on a load of Herefords bred on the SMS Ranch in Texas. it was not until his return home from the war in 1919 that he really buckled down to it, and

As a Man and as a Stockman

then proceeded to make his name and that of his farm familiar in livestock circles across America. Sensitive and modest, he tended to disparage the value and significance of his later accomplishments, but there was little feeling that this was in order among the hard-handed cattlemen who were his contemporaries, or among the livestock-industry spokesmen of the time. Which was not at all surprising in view of the fact that by the end of the 1920's carloads of Casement-fed steers had dressed out at up to 64.59 percent, a level 10 percent above the general run of beeves going to slaughter at that time, according to market observers.

DeWitt C. Wing, managing editor of *The Breeder's Gazette*, leading livestock periodical of its era, wrote of him in the middle 1920's in these words:

"Dan Casement is a militant, fearless, outspoken and dynamic personality, eager to do things on his farm and ranch that would test the mettle of strong men in their prime. He glories in feats of muscle and mind. He budgets his livestock and farming business with thoroughness, neatness and pride. He works hard and zestfully in his feedlots, on the ranch and in his office-den, and loafs a bit with a kind of soldierly ease, dignity and wantonness. He proceeds according to a definite plan, in the manner of an architect.

"Dan is an artist in motive and method. He understands the known principles of heredity and applies them to his Herefords, Quarter

CASEMENT OF JUNIATA

Horses, hogs and dairy stock. But 'fancy' points, with little bearing upon economic usefulness under his conditions, never enter into his plans.

"His chief contribution to the animal industry's stock of knowledge and inspiration is that of an artist in selecting, feeding, fitting, showing and selling fat animals, Hereford steers in particular. Extended experience on a broad scale has trained his mind and hand to select calves that will feed out well . . . When the calves go into his feedlots he not only feeds them well but personally curries and brushes and associates with them until he knows them as individuals, when his voice and presence seem to increase the efficiency of their rations." On occasion, as a long-time friend once pointed out, he might even give them "a dose of Kipling" by reciting a few lines of that famous poet's "Alnaschar and the Oxen," as for example,

"See those shoulders, guess that heartgirth,
  praise those loins, admire those hips,
And the tail set low for flesh to make above!"

It was not at all uncommon for Casement to talk to his animals.

This exemplified a native instinct which probably more than anything else led to Casement's success with livestock. He once wrote that the first impact on his life by a cow critter came when he was tossed sky-high by a heifer when he was four years old. "I learned about heifers from her," Casement once said, adding: "Having been thus catapulted into the livestock in-

As a Man and as a Stockman

dustry, I accepted my fate and my interest has never flagged."

Since his formal education had been in other directions, he cultivated this natural liking for animals by his daily contacts in their feeding and care until his cattle and horses, in particular, virtually developed as personalities in his sight. He almost resented persons who approached them in a strictly impersonal manner. The depth of his feeling for his livestock is indicated by these Casement words:

"The technique of breeding and caring for animals fails, when all is said, to interest me as much as the companionship of it. If a man does not feel joy when he stands in his own feedlot with his own cattle begging for his attention, then he should find some occupation other than livestock tending."

He reiterated on another occasion the logic which so often directed his steps toward a Juniata feedlot where a recently arrived bunch of calves awaited his coming. He wrote, after a session with these youngsters:

"Armed with Scotch comb and rice-root brush, I curried vigorously the arm-pit of one of the 11 calves that I had laboriously gentled, while his 10 tractable companions crowded around and jogged my elbow in their newly acquired eagerness for my ministrations . . . I always like to gentle at least some of my calves when they start on feed or shortly thereafter. The presence of these friendly ones tends to make the others in the lot more quiet and con-

CASEMENT OF JUNIATA

sequently 'better-doers.' Of course, if prize calves from the feeder cattle shows are bought they are already domesticated and thoroughly broken to the feel of the comb and brush. But practically all of my calves have come straight from the range."

So it is apparent that there was a sound practical reason as well as a streak of civilized sentiment behind those feedlot rounds which Casement continued to make regularly until well past his 80th milestone.

He also found frequent trips to the range country stimulating and educational. He rode the pastures of leading commercial cattle ranches picking prospects to fit and feed for carlot fed-steer competition. Details of such activity were related in his latter years in these words:

"For nearly 40 years whenever I discovered that calves in a certain brand had attained show quality I tried to obtain a sample of that brand. In furthering this purpose I bought calves all the way from the cut-over lands of Michigan to the mountains in the Mexican state of Durango, and at many points between.

"I began, I believe, with W. J. Tod's excellent Crosselles from the cañon of the Cimarron in northeastern New Mexico. For several years I prowled out LS's on their old Panhandle range across the Canadian from Tascosa. I had two crops of the LC's from the Baca Grant in Colorado's San Luis Valley. I have had the CS's and the WS's from New Mexico's Cimarron

As a Man and as a Stockman

country; W. B. Mitchell's calves from Marfa, Texas; the Myers cattle from Evanston, Wyoming; Jeff Thompson's from Hereford, Texas; the T-Bones from W. A. Braiden at Antonito, Colorado; the XI's from Plains, Kansas; the HT's from Michigan, and Bull Heads from the American Ranch of Wallis Huidekoper, in Montana." [Huidekoper and Casement met and became friends when the former was a student at the University of Pennsylvania while Casement was attending Columbia.]

In continuing these 1947 recollections of cattle which he had fed in the Juniata lots, Casement went on:

"I fed Matadors for several years, selecting them on the company's Panhandle range. Once I had the Raymond Bell calves from his *Hacienda de Atotonilco* at Yerbanis, Durango, in old Mexico, and once the Marshall Peavey's from Colorado. I have had Gus Dwinell's yearlings from Colorado's North Park, Gage yearlings from Alpine, Texas, and Captain J. B. Gillett's from the Texas Big Bend. For the last three years I have fed calves from the Red River Valley Company—formal name of the Bell Ranch in northeastern New Mexico—and this year I have Mill Iron steers from Wellington, Texas."

There were light-hearted moments in Casement's feedlot visits, which added to the joy he found there. For example, after writing the above, he continued:

"My interest at present is centered mainly on the smaller 60 calves that are destined for the

CASEMENT OF JUNIATA

long feed in drylot. It is with these that I am now cultivating an intimacy. Yesterday I saw one of my bovine friends cut a caper entirely new in my experience. Apparently the brisk wintry air had stimulated his joy of life until it *had* to find expression. He was actually chasing his tail like a cat at play, with the inevitable result that he became dizzy and collapsed."

But, at the same time, as was invariable with him, Casement did not overlook the main goal as he worked with the steers in his lots. "As I have often insisted," he observed, "the true value of a bull is demonstrated only by the quality of beef yielded by his unsexed male progeny. In breeding beef it is on this basis only that the relationship between cause and effect can logically be traced. The butcher's block provides the answer to the cowman's $64 question."

Dan Casement saw his first Hereford bull in 1884 when, as a 16-year-old boy, he visited the Kansas farm with his father who had hired a manager to run it for him. Years later he expressed the thought that this bull, which was bred in Ohio and was recorded in the second volume of the American Hereford Herd Book, must have been among the first Herefords to venture west of the Missouri River. In any case, this breed, then relatively new in America, impressed itself upon the consciousness of the boy to the extent that never in the almost 70 Casement years that followed were Juniata's acres without their Whitefaces.

As a Man and as a Stockman

Casement personally bought his first purebred Hereford bulls for the Unaweep ranch at Denver in 1904. He got his first registered Herefords for the Kansas farm in 1912 from the already prominent herd of Robert H. Hazlett at El Dorado, Kansas, which was destined to become one of the Hereford breed's foremost. Four of the five cows with which he began were bred in the Hazlett herd, and some 30 years later Casement recounted that all of the approximately 1,000 purebred Herefords dropped on Juniata Farm during those three decades were direct descendants of that quintette.

In discussing the first herd bull in Juniata Farm's registered herd, Casement pointed to his selection of Hal Donald from the herd of Jack Cudahy, of the meat-packing family, located just south of Kansas City. In typically Casement narrative he described Hal Donald's hide as "the mellowest I ever touched and as elastic as the rubber in a lady's garter." Then he went on to say, again characteristically, that his "trade with Jack was convivially closed at the old Baltimore Hotel Bar" in Kansas City.

Casement's native astuteness in cattle dealings was exemplified in his purchase of females with which to start the purebred herd on the Unaweep ranch. He went to Gypsum, Colorado, to see Frank Doll's cows, which he later described as having been "big, thick and smooth." His story follows:

"I coveted some of them but realized that I was badly handicapped for intelligent selection

CASEMENT OF JUNIATA

by my ignorance of bloodlines. My only asset was the knowledge that a wide, short head was regarded as the best index to great character in a bull. In Frank's corral I stumbled over a bull's skull which at once arrested my attention.

"Said I, 'Who the hell is that?'

" 'That,' said Frank, '*was* Tempter. He got in a fight with another of my bulls, Pikes Peak, drank too much water when he was all het up, foundered and busted.'

"Assuming a poker face to mask my eagerness, I inquired, 'Have you any of his daughters?'

" 'Sure,' Frank said, 'plenty of them.' "

Casement went on to say that he thus acquired a bunch of excellent cows that justified his intuition regarding their sire's skull. "They furnished an indestructible foundation for the building of the Unaweep herd," he concluded.

As a point of historical interest, Casement scarcely could have done better had he been fully conversant with Hereford bloodlines, because Tempter's ancestry ranked among the breed's best. The bull himself was bred by K. B. Armour of Kansas City, who had been president of the Armour packing company around the turn of the century, while both his sire and dam were close-up descendants of Anxiety 4th, a bull which even yet ranks as the greatest sire in the history of the Hereford breed in America.

Although there is much, much more that might be written about Casement as a cattle breeder, one more example should suffice in indicating the sharpness of his eye. Having suc-

As a Man and as a Stockman

cessfully used bulls of bloodlines featured in the herd of Mousel Bros. at Cambridge, Nebraska, he made it a point to head for the Mousel pen in the yards at the Denver stock show in January, 1929. There, he found a set of 20 bull calves, mostly if not entirely by a sire named Advance Domino, which, he later said, at once won his favor. "They were so uniform in quality, character, and development that there was, in my opinion, little choice between them," said the Kansan in reflection many years later. Coming upon R. D. (Bob) Mousel, who, with his younger brother Henry, owned the bulls, Casement inquired as to the price being asked for his own pick of the five tops in the group. When Mousel set the figure at $2,000, Casement quickly countered by asking Mousel what he would take for a set of five, with the Nebraskan to do the picking. "He offered them for $1,700," Casement recalled, "and I promptly accepted."

He had sized up the bunch correctly for there actually was very little difference between the individual calves in the group. The five picked as the tail-ends of the lot turned out to include Advance Domino 7th and Advance Domino 11th, which had long and distinguished careers in the Casement herds.

Casement's philosophy, as applicable to cattle breeding, was fairly well set forth by these words in his "Random Recollections," a delightful series of reminiscent articles which he wrote during the 1940's for the *American Here-*

CASEMENT OF JUNIATA

ford Journal and which later were reprinted in a small book, under the same title, which has become exceedingly rare:

"I hold firmly to the conviction that, in producing steer calves for the feedlot and heifers for brood cows, real success accrues only to those who are in complete harmony with one fundamental and vital truth, viz., that the human factor in the undertaking is actually a junior partner with Nature herself. I maintain also that this truth applies even more forcefully to the breeders of the bulls which beget these calves. As a matter of fact, it applies to all kinds of productive effort. No form of real wealth can be produced save the process is conducted in strict conformity with natural laws, and in all production we are absolutely beholden to Nature for our supplies of raw material. . . .

"The cowman needs above all unimpeachable integrity not only in selecting the ancestry of his product, but more especially in fabricating it once it is alive and actually in his possession. Since his partner is omnipotent and omniscient he cannot cheat her by any conceivable device his cunning can conjure, but he can play literal hell with himself if he stupidly tries to do so."

While he reveled in the companionship of friends who congregated in the stock show arenas to witness the judging of the individual classes of breeding cattle, it was his firm conviction that "the real show," as he put it, "was in the yards" where the carloads of fat cattle, feeder cattle and, at Denver, bulls in serviceable condition were the centers of interest. "I shall always hold that the true significance of the stock show abides there," he said. He was an outspoken foe of what he considered to be overfitting of breeding cattle for show purposes, and made this observation many years ago: "Beef of superlative quality is made most advantageously on the frames of immature unsexed male animals. Foolishly to attempt a demonstration of this process upon the body of the bull himself, as is the almost universal practice in fitting for show and sale, is a crime against nature and pitifully defeats the purpose for which a bull is intended."

So it was as a feeder and exhibitor of fat and feeder cattle that Dan Casement not only achieved fame in the livestock industry but also gained a spectacular kind of stature in the eyes of the general public. It is the opinion of most of those who are well informed that no man of the soil or the grasslands, before him or since, ever came close to matching him in that respect.

After his American Royal debut in 1907, Casement's fat carloads first appeared at Chicago's International show in 1912 and at the Denver stock show in 1913. During his career as an exhibitor, which spanned 45 years, he showed winning carloads of Hereford steers all the way from Baltimore in the East to Los Angeles and Portland in the West, and at Fort Worth to the South. In 1945 an inventory of ribbons still in his possession which had been garnered in carlot competition at America's leading livestock shows

CASEMENT OF JUNIATA

totaled 316, and, he added, "some have doubtless been lost and others given away."

He continued to exhibit until his last blue and purple ribbons were won at the American Royal in October, 1952, when it was conservatively estimated that his entries had accounted for at least 350 ribbons during his long and brilliant career. The carload of feeder calves on that occasion was bred at Juniata, as were most of the Casement stock-show entries during his later years, were sired by bulls from the Mill Iron Ranches in the Texas Panhandle, which had been his prime source of sires in more recent times, and the calves were tended for both the judging and the auction sale which followed by the then 84-year-old owner. He was honored in the ring before his calves were sold, as the crowd arose as one in a fine ovation. With customary eloquence he expressed his deep appreciation of the friendship of so many fellow stockmen and prophetically said, "This may be my last American Royal." It was. Although it was known only to a few, including himself, he had been fatally stricken.

Although customarily seen in the battered clothes of a cattleman while working with animals at Juniata, he took delight in wearing the flamboyant and picturesque attire which became his virtual trademark at the stock shows. With glasses attached to a wide black ribbon and with pipe almost ever-present, he was the picture of a legend when he stepped out of his work clothes and donned the colorful garb which marked him

As a Man and as a Stockman

in public as the individualist that he was. His own description is better than that of anyone else:

"Very soon now," he wrote in the fall of 1945 a few days before heading for Kansas City and its stock show, "I will delve once more into the depths of my closet and bring forth a suit of black and white shepherd's check, the material for the original of which was brought to me many years ago from Scotland by Murdo and John Mackenzie. I will shake the mothballs out of my brilliant weskit of Stuart plaid, don my red tie and, thus accoutered, sally forth on my annual pilgrimage with a determined purpose to annex another bunch of Royal ribbons."

The Mackenzies, father and son, as all cattlemen of their time knew, were successively at the helm of the Scottish-owned Matador Land & Cattle Company for a period encompassing most of the first half of the 20th century. Quality-bred Herefords from this firm's expansive Texas range early became favorites in Juniata feedlots, and as finished beeves in carload lots won many top awards at major shows for Casement, thereby greatly enhancing the Matador reputation.

In his time with Herefords, it is appropriate to say that not only did Casement make news but also came to be generally considered an elder statesman and something of a celebrity. Of him, *The Kansas City Star* once stated that "the name of Dan Casement is probably familiar to more cattlemen in America than that of any other man

CASEMENT OF JUNIATA

in the livestock business." While he did not in any overtly conspicuous manner cultivate such recognition, except perhaps by wearing his distinctive attire on special occasions, neither did he discourage it.

And, too, he did seem modestly to enjoy the attention drawn by such things as the so-called Casement-clip—a close clipping of the hair on the heads of his show loads which seemed to make them appear even more beefy by contrast—a practice which soon came to be widely imitated. Another Casement touch which got considerable publicity in his later years was that of dipping the tails of winning carloads of feeder calves in purple paint. Attributing this innovation to his son Jack, he said as he gazed appreciatively at his purple-tailed grand champion carload at the 1946 American Royal: "They seem to look prettier when the sunlight hits their tails." Besides which, the color matched that of the grand championship ribbon and was a plainly visible reminder to all who passed by of their victory.

A Casement Family Album

This reproduction of a large oil painting, now the property of Mary Casement Furlong, shows Dan D. Casement, then four years old, in white dress and blue shoes, holding the family dog, "Flash," and beside him his six-year-old brother, John Frank Casement, in red velvet suit. The flagdrum at lower left was sketched in recognition of their father's military exploits. The boys were posed on the front porch of the new house at Jennings Place. John Frank died at the age of 19, and another brother, Charles Jennings Casement, lived only until his fifth year.

Mr. and Mrs. John Stephen Casement, parents of Dan D. Casement, as taken from a daguerreotype made shortly after their marriage in 1857. This picture of them, says their granddaughter, Mary Casement Furlong, is "my favorite."

Charles Clement Jennings, maternal grandfather of Dan D. Casement on whose farm, Jennings Place, Casement was born, and where his boyhood was spent.

The father of Dan D. Casement, known across America at the height of his military and railroad building career as "General Jack."

This picture of Mrs. J. S. Casement was taken on the porch at Jennings Place in the summer of 1917, just after her son Dan had enlisted for service in World War I. "Grandmama remained cool and knitted for her war effort," says her granddaughter, Mary Casement Furlong.

General J. S. Casement with his construction train at the "end of track." He and his brother Dan, for whom the latter-day Dan D. Casement was named, comprised the firm of Casement Brothers, and laid all of the track and did most of the grading for the Union Pacific Railroad from Omaha to Promontory Point, Utah. The photographer's traveling darkroom is at the right.—A. J. Russell photo from Union Pacific Railroad Museum Collection, Omaha, Nebraska.

Up and down the line on horseback the General went as he maintained a close watch on the progress of construction of the Union Pacific Railroad. Although he could stretch out to not much more than 5 ft. 4 in. in height, he turned out to be 10 feet tall when it came to handling men and executing plans in railway construction.—A. J. Russell photo from the Union Pacific Railroad Museum Collection, Omaha, Nebraska.

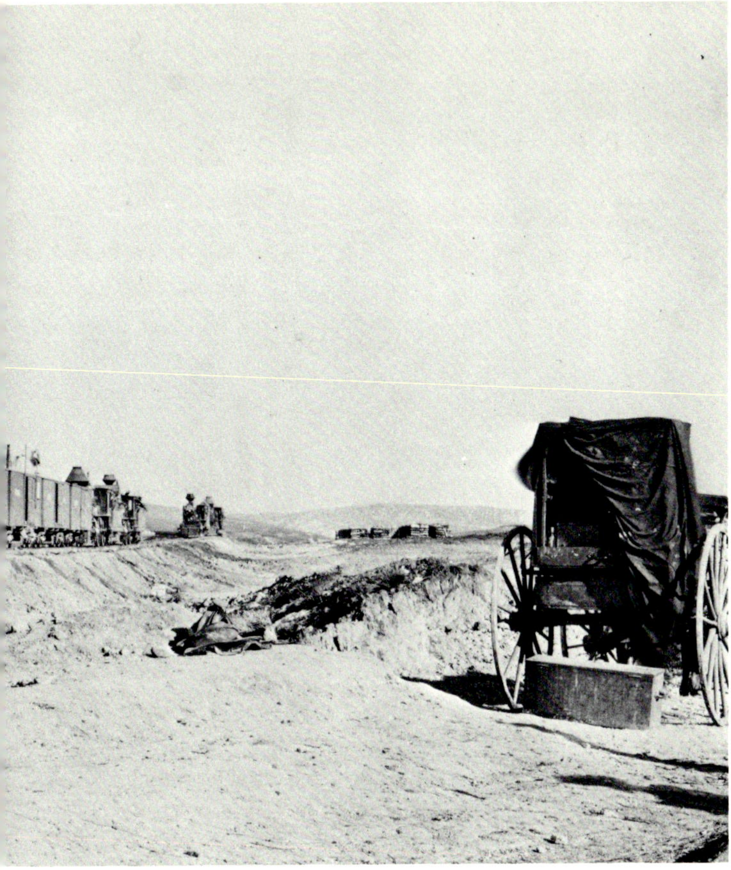

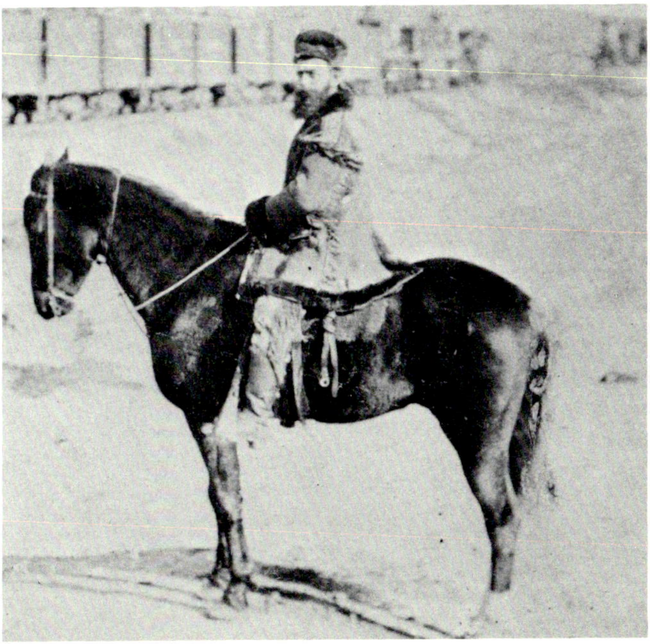

The Casement House and Jennings Place

The original pen-and-ink sketch of Jennings Place from which this reproduction was made was done shortly after construction of the new brick home and other buildings was completed in 1871. Dan Casement was born in the original story-and-a-half frame house that stood on what became the front yard of the new home, and was three years old when the family moved in. Situated just east of Painesville, Ohio, the place has a long and rich heritage which vividly recalls the magnificent Victorian era.

When Charles Clement Jennings' daughter, Frances, was being wed to General "Jack" Casement, he decided to build them a home and present it to them as a belated wedding present. The house took three years to build, and when it was completed it had cost Mr. Jennings $75,000. It has been estimated that such a house 100 years later, if it were possible to duplicate, might cost more than a half-million dollars.

All of the partitions in the house run from the basement floor to the attic ceiling, and all of the wood which was used was cut from the timber which was on the 300-acre Jennings farm. The house had 24 rooms and nine gas fireplaces, fuel for which was piped to the house, not only for heating but also for lighting and cooking, from five natural gas wells dug on the farm.

Stables, barns and lamps on posts along the driveway were similarly lighted. Top craftsmen of the times made decorative wood carvings for the house, and artists came from New York to paint festoons of cupids and flowers on the ceiling of the drawing room and also in the vestibule. A unique feature about the house is that it had its own air-conditioning system. In constructing the house, separate wooden duct work was built into the walls, and through them passed cool air which was drawn in from the outside under the porches.

The property, although purchased in 1953 by Mr. and Mrs. Robert W. Sidley, is still known as The Casement House, and continues to reflect the rich heritage and tradition of the old Western Reserve. In the years since 1840, the property has had only four owners: Charles C. Jennings, General J. S. Casement, Dan D. Casement, and the Sidleys, who bought it after Dan Casement's death.

Significant features of the accompanying sketch, reproduced from the original which is in the possession of Mary

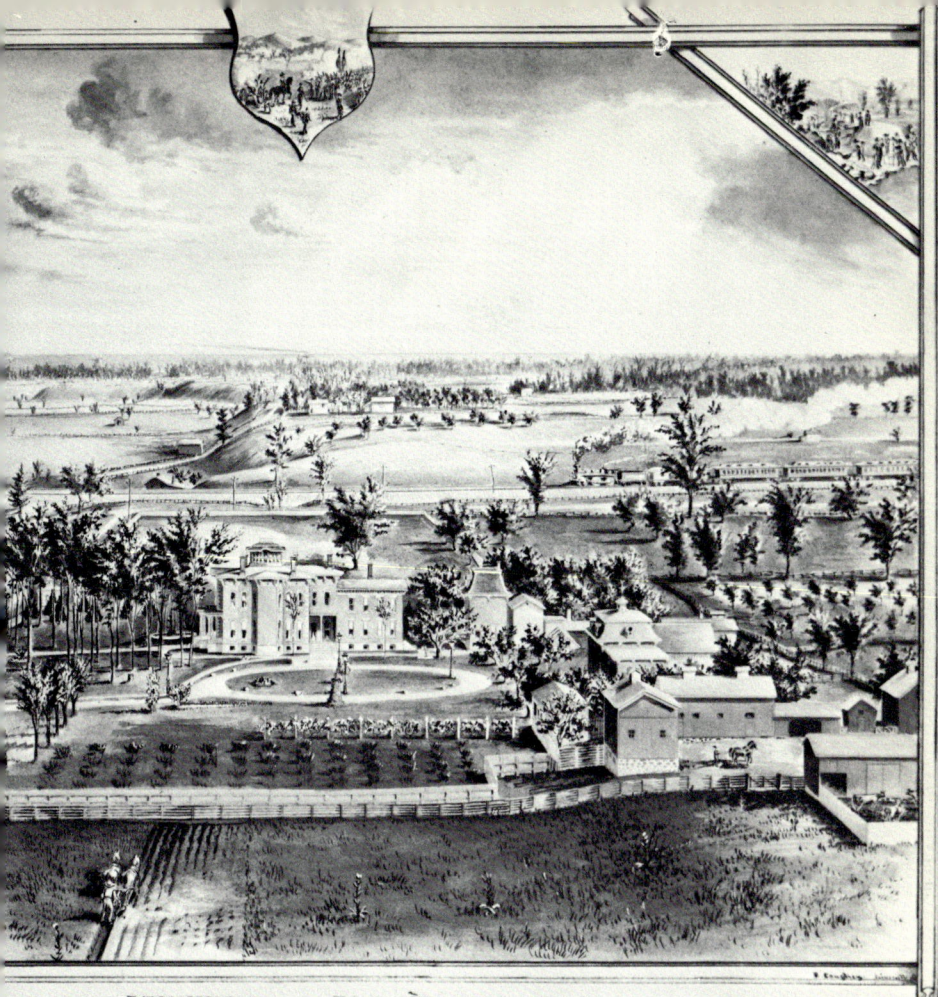

JENNINGS PLACE.
OF GEN. J. S. CASEMENT.   PAINESVILLE O.

Casement Furlong, are pointed out: Please note that railroad rails frame the scene, and that the three small pictures across the top depict highlights in the General's career—one of laying railroad track, one of building a bridge, and the one in the middle showing General Casement in battle. Observe that two trains are about to meet across the background, and that smoke from two Lake Erie steamships is visible on beyond. A carriage is in the driveway and a farm worker is plowing in the foreground.

The main house and nearby carriage house, stable and ice house were of brick construction. The farm buildings were built of wood, and mostly predated the new house. The General later built a porte-cochere over the side porch and enlarged the front porch (small picture). The buildings are intact except the cow barn, at front right, which burned in 1974. It was in this barn that Mr. Jennings during the Civil War kept runaway slaves in the loft and at night took them to the lake shore to board boats that carried them north to Canada.

Above is Unaweep Cañon, looking toward Thimble Rock. Bunk house, barns, sheds and corrals of the Unaweep ranch are at the left, while the little red ranch house is at right. A large reservoir in an alfalfa field does not show in this picture.

Apple trees blooming in spring-time are seen in the picture at upper right, at side of the little red ranch house on the Unaweep. The red had aged to a faded pink by the time this picture was taken. The small pond in the front yard was supplied by Fall Creek, at rear. All furniture was of rustic vintage. Wall paper remained intact, but was faded. Gibson Girl pictures were displayed on the walls, along with a series of Charles M. Russell western prints. Lariats and giant Stetsons were hung on deer feet or antlers, utility thus joining ornamentation.

Thinking back to happy days at the little red house on the Unaweep, Mary Casement Furlong recently wrote: "No one ever locked a door as there were no keys. On our return yearly, everything was in place. Only mice and pack rats intruded, which meant the 'women folk' had to wash all china, pots and pans, and put clean papers on the shelves and in bureau drawers. Over doorways in living room, written by D. D. C., were quotations from Ovid or Horace. One I remember was the equivalent of 'Far from the maddening crowd,' in Latin. Wish I had copied them. The Gibson Girls never faded and were intact from year to year. Nothing changed. Mother did make new

The Piñon Mesa roundup crew in western Colorado in 1892. The small arrow near the center points to Dan Casement, then a 24-year-old rancher on the Unaweep range. On his right was Henry Knowles and on his left sat Sam Pollock. "Such were the good men and true," Casement once wrote, "with whom I rode the range."

lamp shades, and brought curtains for windows, and linen and blankets as needed. All lamps used oil.

A homemade book case filled with novels of the '90s was at one end of the living room. In the room was a large stove, a couch, an old secretary desk, two Victorian tables, a rocker and chairs covered with Hereford hide in their seats and backs, and on the floor a bearskin rug and three or four others. An oak dining table and chairs and a built-in china closet were in the dining room. There was a large kitchen and pantry, with a wood stove and reservoir. There was no running water in the house, so water was carried by bucket from Fall Creek. A large 'keeping' room was off the kitchen, in which to hang venison, bacon, etc. There was a cellar in which to keep milk, and a privy was at the rear.

Dan D. Casement astride the frisky Concho Colonel on the Unaweep range around 1911. Son Jack, 3, was near, and his sister, Frances, was at the picture's edge.

Dan Casement spent the first three winters after returning from Costa Rica in Washington, D. C., where Mrs. Casement's mother lived and where this picture was taken. There he lived a life of ease, behaving "more or less like a playboy," as he was later to say.

The girl Dan Casement was to marry was Miss Mary Olivia Thornburgh. So smitten was he that three nights after meeting her he asked her to marry him, and she consented. But before they were married he spent many a lonely evening on the Unaweep writing letters to her. They were wed at All Angels' Church in New York City on December 1, 1897, and sailed three days later on the Alene for Costa Rica, and the railroad construction project.

Dan Casement and his wife, Olivia, as she was known to friends, pictured in San Jose, Costa Rica, as they gazed enrapturedly at their first-born, daughter Mary Eliza, then seven weeks old.

Dan D. Casement and his father, General Jack Casement, spent six busy and happy years in Costa Rica building the Ferrocarril al Pacifico, a railroad extending from that Central American country's capital city of San Jose to the Pacific Ocean. The youthful Dan, newly married, and his bride, along with his father and mother, entered freely into the social life of the San Jose region at the turn of the century. The group pictured below in the garden of the Casement residence, La Sabaña, on the outskirts of the city, shows, standing left to right: Edmund A. Osborne; Mrs. Daniel Casement, whose husband was the General's brother; Mrs. Dan D. Casement; Miss Lilly de Jongh, a girl from Holland who later became a prominent author; General Casement, and Dan D. Casement. Seated, from left, were: Mrs. John S. (Jack) Casement; Mary Eliza, daughter of Mr. and Mrs. Dan Casement, who was born in Costa Rica, tending the family's Great Dane, and Mrs. T. T. Thornburgh, mother of Mrs. Dan D. Casement, with small dog in her lap.

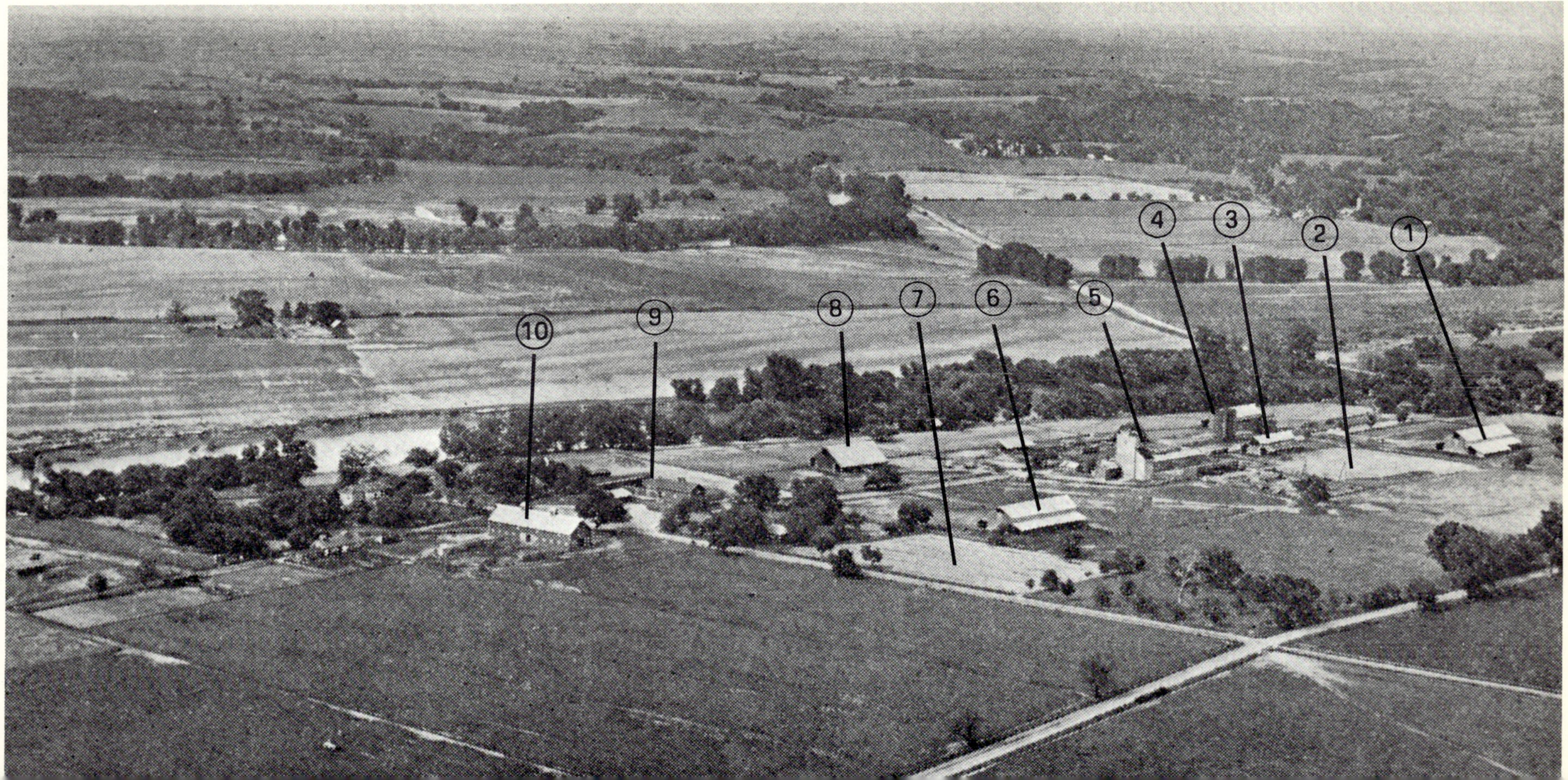

1
2
3
4
5
6
7
8
9
10

The cluster of buildings above, in their owner's heyday painted red with white trim, comprised the headquarters of Dan Casement's Juniata Farm. They were situated at one end of the section of crop and meadow land which constituted the nucleus of Juniata operations. Identifications of the various facilities, as provided by Casement's long-time Manhattan friend, Orville B. Burtis, Sr., follow: 1, 3, 6 and 8, cattle barns with hay down the middle and grain self-feeders at the sides of the sheds of each, which were kept filled with grain with the help of the Suffolk stud, Baby Charlee, hitched to the feed wagon; 2, a small lot where show loads of feeder calves often were put before shipping out; 4, silos; 5, grain elevator and feed mill; 7, a small paddock into which the sows frequently were placed as their litters of pigs were weaned; 9, an all-purpose barn, and 10, the horse barn. The Big Blue River flows among the trees immediately beyond the buildings.

This pen and ink sketch of Juniata Farm headquarters depicts it as it was in the early years of Casement ownership. The sketch is thought to have been made around 1880 before General Casement transferred ownership to his son, Dan. The line at bottom was added later.

Dan Casement was a firm believer in "horse power" in the operation of his farm and feedlots. — Gene Guerrant photo.

This big house at 610 Humboldt Street in Manhattan was the Casement family home for many years until Dan Casement's death in 1953. Casement hospitality was generously extended here to generations of stockmen and leaders from other walks of life who found their host one of the most unusual farmer-stockmen of all time. Likewise, the attractive garden behind the house was a frequent gathering place when Casement guests assembled.

Dan Casement "toasts" the "Happy Holidays" with this picture of himself and the almost ever-present coal-black Cocker Spaniel, "Nicodemus." This picture illustrated his greeting card which was sent to friends to mark the 1950-51 Holiday Season. — Gene Guerrant photo.

The living room at 610 Humboldt Street, extending on into the drawing room, where Casement was a gracious host.

This picture of Dan Casement, in his stock-show garb, was taken in Kansas City not long after he moved from Colorado Springs to become a full-time Kansas resident and the guiding hand in Juniata's success.

Dan Casement, at left, below, and John Collister, Juniata foreman, at halter, displayed a Juniata steer to a visitor.

Juniata Farm frequently was a gathering place for notables in the agricultural and livestock field. In the long-ago assembly below, left to right, were: Joseph H. Mercer, state livestock sanitary commissioner of Kansas and secretary-treasurer of the Kansas Live Stock Association; Dan D. Casement; W. M. Jardine, former president of Kansas State Agricultural College and subsequently United States Secretary of Agriculture; United States Senator Arthur Capper; Robert H. Hazlett, famed Kansas breeder of registered Herefords; unidentified, and Dr. F. D. Farrell, president of Kansas State Agricultural College.

Finished prime beef on the hoof like the grand champion carload (above) at the 1931 American Royal show brought wide prominence to Casement and Juniata. These steers were bred in the Texas Panhandle by the Matador Land & Cattle Company.

Cold weather never kept Casement from giving personal attention to his cattle, at home or away.

This Casement-fed grand champion carload of steers in the 1930's was bred at the American Ranch of his good friend, Wallis Huidekoper, at Two Dot, Montana.

Last great favorite among Casement's Quarter Horse mounts was The Deuce, "a replica of his sire," Casement once wrote, adding: "He is my pride and joy. I often tell him so." — Gene Guerrant photo.

Dan Casement in the Bluestem Hills makes friends with a member of his Quarter Horse contingent.

Whether populated by Herefords or Quarter Horses, as in the instance below, this scene was typical of Casement's Bluestem pastureland some five miles north of Juniata headquarters. The setting of this picture of a band of Quarter Horse brood mares and The Deuce was the first pasture on Cedar Creek.

When on the job at Juniata, Dan Casement frequently spent time, as he put it, "currying favor with calves." There was a functional purpose in this, as well as genuine pleasure for both the steer and owner. "The presence of the friendly calves," their owner once observed, "tends to make the others in the lot more quiet, and consequently 'better-doers'."

Most of the prize-winning Casement carloads of feeder calves and finished beeves during the later years of his Juniata operations were products of his own commercial Hereford breeding herd. That they displayed abundant evidence not only of good feeding but also good breeding was apparent in this load, which the sign indicates was "Bred, Fed & Shown by Dan D. Casement."

This set of steers found the air brisk on a winter day at Juniata. They had been handled by the deferred feeding method and were headed for Denver's National Western Stock Show.

*When he was at home Dan Casement seldom missed a day not only in looking at but actually studying the individual animals in his feedlots.
—Gene Guerrant photo.*

Dan Casement in this salering at the Chicago Feeder Cattle Show in 1946 faced a multitude of bidders who rendered the final verdict on the value of this winning carload of Juniata feeder calves. That verdict in this instance was a price of $42.50 per cwt. Standing at lower left are, first, W. E. Ogilvie, manager of the International show, and next, R. J. Kinzer, long-time secretary of the American Hereford Association.

At the 1946 American Royal Livestock Show in Kansas City, Dan Casement visited in the judging arena with Don Ornduff, who had become editor of the American Hereford Journal *a year or two earlier. He enjoyed visiting with friends, old and new, as he watched the judging of the individual breeding classes, in which he was never an exhibitor.*

Casement's agricultural interests were as broad as the entire field of agriculture. Here he stood in a ripening field of wheat with John Parker, of the Kansas Wheat Improvement Association.

Dan Casement could "dress up" when the circumstances made it appropriate. Here he stood on the lawn at his home in Manhattan with Charles A. (Tot) Otis, of Cleveland, along with a "Warrior of The Black Watch," where they had just witnessed the marriage in the Casement house of two young friends. Casement once described Otis as "my oldest friend and early ranch partner of the Unaweep days, who has stood by me on all the great occasions of my life."

Dan Casement, left, seated in an Albuquerque hotel lobby during the 1949 convention of the New Mexico Cattle Growers' Association. With him was Capt. Burton C. Mossman, one-time superintendent of the famous Hash Knife range outfit in northern Arizona, owned by the Aztec Land & Cattle Company, and later the leader of the equally famous Arizona Rangers who brought law to the Arizona Territory. Casement "knew them all" in the heyday of his career.—New Mexico Stockman photo.

In his 45th year as an exhibitor, Dan Casement at the 1952 American Royal Livestock Show in Kansas City stood among the calves which comprised his final champion carload of Hereford feeder cattle.

Casement was honored in 1952 at the annual meeting and banquet of the American Hereford Association for his contributions to the field of animal agriculture. In the picture below, he stood between Jess C. Andrew, West Point, Indiana, then president of the International Live Stock Exposition, at left, making a presentation, and Roy R. Largent, Merkel, Texas, retiring president of the Hereford organization. Jack Turner, then American Hereford Association secretary, looked on at the left.

This final family picture of Dan Casement, his three children and their spouses, and one granddaughter, was taken at the family home in Manhattan in January, 1952. In the top row with Dan Casement is his son Jack, and Jack's daughter, Julie, from Sterling, Colorado. Seated, left to right, are: Harold Furlong, son-in-law, from Painesville, Ohio; Mrs. Jack (Xenia) Casement; Mrs. Harold (Mary) Furlong, elder daughter; Mrs. Donald (Frances) Dorn, his other daughter, from Mexico City, Mexico, and her husband, Donald Dorn. Jack holds Valentina, the tan Dachshund which belonged to Frances.

DAN D. CASEMENT
MANHATTAN, KANSAS

JUNIATA FARM

HEREFORD AND AYRSHIRE CATTLE
SUFFOLK-PUNCH AND AMERICAN QUARTER HORSES
DUROC AND POLAND CHINA HOGS
HAMPSHIRE SHEEP

February 17, 1952.

DAN D. CASEMENT
MANHATTAN, KANSAS

JUNIATA FARM

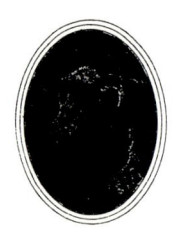

HEREFORD AND AYRSHIRE CATTLE
SUFFOLK-PUNCH AND AMERICAN QUARTER HORSES
DUROC AND POLAND CHINA HOGS
HAMPSHIRE SHEEP

March 3, 1953

Dear Mr. Ornduff:

Herewith is a final Juniata Jotting — and perhaps the longest one I have sent in. It deals with a controversial subject and you may not care to print it. However, I do not see how it could reasonably give offense to your patrons, because, it is, after all, the opinion of myself alone and I shall not be embarrassed or disturbed by whatever criticism it evokes.

If in your judgement it would be unwise for the Journal to print it, please return it promptly.

Very sincerely,

Dan Casement

Dan D. Casement

P. S. I write in bed, my other infirmities having been complicated by the flu since I returned from Denver.

Casement letterheads were distinctive in that they were printed entirely in a rather bright red color. Only a change of picture was observed in the Casement letterhead over a period of a good many years of correspondence with Juniata's owner. The upper picture showed Dan Casement astride The Deuce, while the other illustration was of one of his herd bulls. The letter reproduced from the latter letterhead, dated March 3, 1953, must have been one of his last. Four days later he died.

Part II

Next to his Herefords, Casement had extraordinary affection for horses. This dated back, he once said, to his Ohio boyhood when he had an Indian pony named Punce. He was among the earliest enthusiasts for Quarter Horses, acquiring his first stud, Concho Colonel, in 1911 from William Anson, of Head of the River Ranch, Christoval, Texas. Always thereafter this breed figured prominently in his "way of life" and in 1940 he became one of the organizers of the American Quarter Horse Association, which he later served as an officer. It was at about this same time, when he was 72 years old, that he won a cutting horse contest at the American Royal, "sitting loose in the saddle with a slack rein," as it was reported at the time.

Robert M. Denhardt, prominent Quarter Horse historian, who knew Casement well for many years, later wrote of him: "In many ways Dan D. Casement was the greatest American I have ever met, or ever expect to meet."

Casement's sentimental attachment to his equine friends is highlighted by the following from his "Random Recollections" series: "For nearly two score years three grand horses—three generations of the same blood, Concho Colonel, Balleymooney and The Deuce—shared my life most intimately. In work and in play they were my faithful friends. In my lighter hours they added greatly to my gaiety and happiness; in times of trouble, I turned to them for consolation. Happy indeed is the man who lives with mutual respect and affection in close comradeship with a noble horse."

How he relished "working" cattle astride an able Quarter Horse, as befitted a cowman of the old school! Back in the middle 1920's, when Balleymooney was his favorite, he once wrote in appreciation of his mount's skill in gathering a bunch of cows. He said:

"I rode Balleymooney. He was in exceedingly high spirits, and impatient of any display of wilfulness on the part of the cows. Comprehending perfectly that his business is to drive cattle, he is apt, if left to his own devices, to employ hasty and severe measures in the prosecution of his job. Repeatedly, with head extended low and neck curved concavely after the manner of a menacing gander, he rushed at the 'drags' and his teeth came together with a report like a pistol when they slipped from the tailhead of some particularly provoking and stubborn cow."

Not only the Quarter Horse but also the Suffolk drew Casement's admiration, a stallion of the latter breed, Baby Charlee, coming to his farm from the ranch of William Anson at about the same time as Concho Colonel. In addition to his stud duties, one result of which was

As a Man and as a Stockman

a set of mares on which Casement's Quarter Horse stallions sometimes were crossed, Baby Charlee was in harness almost daily in performance of his farm chores. "He served the farm faithfully," Casement once wrote, "hauling the feed wagon practically every day since his arrival . . ."

Casement's deep sense of compassion for his animal friends never was better illustrated than when he wrote of the death of Baby Charlee who, suddenly stricken, was unable to respond to prompt veterinary treatment. "As the end seemed near," Casement wrote, "he followed us from the stable out into the sunshine, and to each of us standing around him he came in turn as if beseeching the help that we were powerless to give. He met death courageously, standing up. We buried him under the trees by the river and set a large headstone at his grave . . . He was handsome, faithful and true."

Dan Casement likewise felt a special brand of companionship with dogs. From the days of Jack and Juno, pointers with which he hunted on the Ohio farm of his boyhood as early as the age of 10, to the time of Nicodemus, the silky black Cocker Spaniel of his final years, whether in town or going to the farm in the ancient Casement "Chevy," he seldom was to be seen without a dog. Nor did the dog have to be highly trained in hunting skill. It seemed to be more a matter of comradeship. For example, after hunting quail with Spot, the pointer of an employee, which he described as "not a finished field dog," he wrote:

"For four short days this season we associated happily—if somewhat profanely—with Spot and his vagaries. My score for the entire season was 15 birds in the bag. One thing puzzles me a little. I'm using the same gun I used 30 years ago but it doesn't seem to bring down so many birds. I find some consolation in the fact that, at 79, I do not feel the same keen delight in taking a life as I did at 19."

The reference above to the Casement Chevy is a reminder that he regarded something as inanimate as an automobile merely as transportation. To a visitor some years earlier, who got into another muddy, much-used car to go with him from town to the farm, he said with an air of exuberance as he patted the steering wheel with one hand: "She's old and battered, but she's got the power. I could jump fences with her if I had to." Which perhaps after all indicates that he did attribute something in the way of personality to this "inanimate" object.

As with hunting, Casement found life-long delight and special zest in the sport of fishing. After starting with bass as a boy in Ohio, he fished as frequently as possible throughout his adult years in the rushing waters of western streams, probably beginning with Fall Creek on the Unaweep ranch. Mary Casement Furlong, the elder of his two daughters and the last survivor of his three children, touched upon this

As a Man and as a Stockman

location in June, 1974, when she wrote of some early outings with her father. She said:

"The little red ranch house on the Unaweep, I was privileged to share with him many October months until my marriage. Here he returned to his youth, shot an occasional deer or mallard for the larder, and 'wet his fly' in Fall Creek. It was marvelous to see him crouched in the willows striving to place the 'Royal Coachman' or 'Gray Gnat' under a certain mossy rock to lure a wily trout."

How this man loved fishing and the accoutrements of the fisherman! Preparing for a trip to test his skill in Colorado's Taylor Creek, above Gunnison, he had been given by his family and two old friends a new fly rod, a "fine Hardy reel," as he described it, and "wonderful stocking waders and felt-soled shoes." "Thus equipped," he later wrote, "I felt like a small boy the day after Christmas."

On this same trip, in the fall of 1947, he went one day to the opening of the famous dispersion sale of Thornton Hereford Ranch near Gunnison. After reveling in the companionship of scores of friends the evening before the sale, he watched the beginning of that historic event, of which occasion he later wrote:

"After a few animals had passed through the ring and Art Thompson's hammer had knocked them down to the tune of some $200,000, I nervously fingered the dollar and six bits in my pants pocket and went fishing." Starting on Ohio Creek, just outside the Thornton fence, he

CASEMENT OF JUNIATA

fished the creek to the Gunnison River, and on down the river to a rendezvous with his longtime Manhattan friend and fellow rancher, Orville Burtis, who met him with the car at a highway bridge.

"It was a strenuous effort for one in his 80th year," Casement later related, "and rather unproductive, but it was a perfect afternoon. The tall cottonwoods along the bank were gloriously golden and a young buck, surprised by my sudden appearance, seemed almost friendly!"

A day or two later, he went fishing again, on a small tributary of the South Platte on Tom McQuaid's ranch in Colorado's South Park. Here his luck was better: "Trout rose eagerly and soon my creel was laden. After brief indulgence of delight in this elysium, I gutted my catch and, as always, left my knife on the spot. I know where it is and hope to go back and recover it next summer."

On another occasion that same year, his luck was not all good. Using waders borrowed from Andy Anderson of the A Bar A in southern Wyoming, he caught his biggest rainbow of the season. Then, in his own words, "I tripped on a boulder and fell full length in the North Platte—but nothing could dampen my delight."

Early the following spring, unable to wait to go where the trout are, he wrote about going fishing on Cedar Creek, in his hill pastures north of Juniata headquarters.

"I am a confirmed fly fisherman," he related, "and, with no trout stream nearer than the

As a Man and as a Stockman

Ozarks or the Rockies, I am forced to exercise my imagination and practice my favorite sport vicariously. Cedar Creek in April often has all the charm of a trout stream. It is clear and abounds in alternate swift water and enticing pools. Following its course through the woods with a two-ounce rod and a Number 12 barbless fly induces a sort of hypnotism that makes the lowly chub look like a trout to me, at least when a nine-incher rises avidly for my lure, as four of them did today."

This extraordinary man was still "goin' fishin' " when well past the age of 83, as shown by this letter, dated September 20, 1951: "I am leaving for Colorado Monday morning for a brief whipping of the stream," he wrote.

Those who might have been inclined—and there were some—to look at Juniata Farm as something of a "hobby" operation were soon set straight. It was Casement's firm conviction, and he had the figures to prove it, that he operated an eminently practical and profitable farming and livestock enterprise. Profit was essential for the continuation of his operations, he emphasized, and to that end he worked diligently, handling every practical detail in connection with both the livestock and the land. Over one continuous period of 37 years, for example, his books at year-end were "in the black" in all but nine years.

He was a keen student and observer as he traveled about, and also cooperated closely with

CASEMENT OF JUNIATA

Kansas State Agricultural College in matters relating to his livestock and to his crops and pastures. Students frequently visited the farm, knowing that those of serious purpose would be welcome, and he participated in farm-and-home-week programs at the college and cooperated on other occasions. He once wrote of responding to a request from the dean of agriculture to address the class of 1925 on the subject of "The Farmer's Ideals," stopping work on what he was doing and reluctantly donning, as he later said, his "citizens' clothes" to appear. Then he hurried home in time to help house the ewes for the night and to de-tail the latest couple of arrivals in the flock.

He was always willing, at home or away, to encourage young men who were genuinely interested in livestock. This point is reinforced by another incident of the middle 1920's, as related nearly a half-century later by L. P. McCann, then coach of the livestock judging team at Colorado A. & M. College, and later a staff member of the American Hereford Association. McCann wrote: "About 1926, I was instrumental in getting Dan Casement as the speaker for our college judging team banquet during the Denver stock show. When Dan was introduced he pushed back from the end of the banquet table, placed both feet on the chair seat, and sat down on the top of the chair's back. From this position he proceeded to deliver one of the most interesting and entertaining banquet speeches I have ever heard."

As a Man and as a Stockman

Likewise he was a popular figure at Fort Riley, the historic military post located not far west of Juniata, which is not surprising in view of his own army background. In its heyday as cavalry headquarters, he found much pleasure in watching the horse shows there, in judging when called upon, and in his contacts with the men with whom he felt a special affinity.

"It warms my heart," he once wrote of those times, "that to many of the fighting men and their brave young wives at Fort Riley I am 'The Deever,'" doubtless in recognition of "Danny Deever," one of the best known characters of Rudyard Kipling, who was one of Casement's favorite authors. At least one of the Casement daughters also frequently and fondly referred to her father as "The Deever."

He also was on call for other occasions at Fort Riley such as, for example, when he addressed the Officer Candidate Class in June of 1951. Citing the young soldiers' traditional "unconquerable spirit" implemented by their "natural endowments as men," he went on to say:

"The love of liberty is instinctive in man and is about the most important and tangible of his moral attributes . . . Human freedom found its firmest if not its first abode on this continent and fortunately was evoked to serve as an ideal in the foundation of this nation."

But that he didn't like what he saw happening in his country is evident in these words ex-

CASEMENT OF JUNIATA

pressed near the conclusion of that Fort Riley talk:

"How strange that men should willingly forfeit a thing more precious than life itself! For the patriot's plea 'Give me liberty or give me death' was not a mere oratorical platitude. It voiced the reality of the noblest and dearest human impulse. The truth is we have allowed our politicians to trade our freedom with its blessed adjuncts of discipline, duty, decency and all manly virtues for a false promise of security such as only a slave would covet or accept."

That was how Dan Casement felt about liberty and freedom.

He had similarly firm convictions about the necessity for productive labor. "Security never produced anything but lassitude," he said in a speech to the Manhattan and Junction City Rotary Clubs. "The keynote in the making of America," he went on, "was its lack of security, with a corresponding abundance of opportunity provided by a completely free enterprise. The do-and-dare spirit, not the paternalistic spirit, made us what we were."

Acknowledging the ills that by the late 1940's had beset America, he still felt that there was some reason for hope if America would "reorganize politically on the pattern set by our original constitution." He urged his fellowmen "to live in harmony with the creative intelligence and in conformity with the immutable laws of Nature. The instinctive moral laws," he went on, "are codified in the Ten Command-

As a Man and as a Stockman

ments. Generations of trial and error have proved their wisdom and established their righteousness. The sin for which we must now suffer is that we have sought to nullify natural laws in catering to man as wholly an economic animal in disregard of his moral attributes."

Dan Casement's agricultural instincts were as deeply implanted, and probably reflected his urgent sense of freedom. His teen-age expression of a desire to farm was a firm commitment from which he did not waver, as was shown by this philosophical comment written after many years of experience on the land:

"It would be a dull mind indeed that would find no stimulation and incentive in the infinite possibilities involved and the engaging problems encountered in the effort to improve domestic animals by breeding. In this field man is working in close partnership with the infinite force that rules all nature. The deeper his insight into its laws, the closer his communion with it, and the fuller his realization of his obligations in this partnership, the more certainly and helpfully will his vocation react on his personality and the more surely will he be rewarded with material success and the personal satisfaction that comes with worthy accomplishment. No business calls for greater skill, keener insight, truer intuitions.

"On the material and technical side its requirements are no less exacting. Fundamentally, the stockman is a manufacturer, producing an article almost universally essential to human life. His job is not only 'to evoke from senseless nothing to conscious something,' to serve as the medium on which he builds and stores his product, but also to produce that product in the most effective and economic manner, and present it for sale in the most attractive form. Unlike the usual manufacturer or builder, he is not fabricating with inanimate materials or building with metal, wood or stone, but with breathing, living organisms of sensibility and personality. That fact infinitely increases the fascination of his business, while complicating the factors that enter into its success or failure.

"His finished product represents the beneficence of nature, allied with his own intelligence, energy and industry. Its value depends largely on the quality and quantity of these personal elements that he expends on its production. To a greater extent than in the case of most other vocations, he is independent of uncertain and disillusioning human contacts. At the hands of Nature herself are bestowed his greatest successes, and from the same source come his hardest reverses. The broader and more versatile his knowledge, the better are his chances. Through ripe experience and wide observation he may discover and develop 'truth and God's own common sense, which is more than knowledge,' and which serve him to better purpose than do treatises on the laws of breeding and nutrition, although familiarity with such subjects may well be a part of his intellectual equipment."

As a Man and as a Stockman

Despite the fact that the cattleman must seek his hard-earned compensation under the strict discipline of the elements, Casement reflected on another occasion upon some of the special skills and aptitudes that distinguish the good cowman and make him most efficient in doing his daily work.

"At the top of these," he wrote, "I would rank a profound comprehension of the instinctive behavior of the native range cow, undefiled by domesticity. The good cowman, in short, must understand the heart of a cow as intimately as a great and good man understands the hearts of men. Thus equipped, he is able, in working or trailing a herd, always to be in the right place at precisely the right time. Reasoning like a cow, one might say, he can anticipate her probable conduct in any set of circumstances and thus beat her to the break with a minimum of effort.

"Undoubtedly great skill in this art was acquired most readily in working the long-horned Spanish cattle of long-ago trail days. Upon the invariable pattern of their impulses the alert rider could place a sure dependence and thereby gain a mastery of his profession, to be handed down by precept and example to his fortunate sons and worthy successors of later generations. The best of modern cowmen owe their proficiency to such sources. A pronounced aptitude for handling cattle may truly be classed as an inherited gift.

"It goes without saying that the good cowman is likewise a good horseman. He is entirely dependent upon his horse in the performance of nearly every important duty incident to his profession . . . He understands and respects the nature of a horse as fully as he comprehends the nature of a cow brute. Horse-sense as well as cow-sense, coupled with the ability to apply them promptly, quietly and confidently to the demands of the work at hand, are unfailing marks of the good cowman.

"In listing the essential personal virtues of the good cowman, integrity must, of course, be placed first, as it must in the character of all men of whatever calling. Common sense, courage, endurance and self-reliance are almost equally important requirements. He must work fearlessly and he must never be a quitter. His self-respect imposes upon him these stern necessities.

"From an esthetic viewpoint, life on farm, ranch and range is unrivaled, surrounded as it is by Nature in all her beauty as well as in her sterner moods, and influenced by the subtle charm of the changing seasons. Contented cattle grazing rich pastures in the foreground of a beautiful landscape form a picture as pleasing to an artist as to their owner, and the panorama of broad and fertile acres touches something deeper in a man than mere pride of possession.

"The farmer, the rancher, the stockman—all those who own and occupy the open country—should nourish and strengthen their attachment to it, and, by broad thinking and sound education, both cultural and technical, fit themselves

As a Man and as a Stockman

and their children to live level with the responsibilities, opportunities and honors that await them."

Combined in Casement was a rare blend of the high idealism this comment indicates and an awareness of the practical necessity of making his operations pay. At the end of one 15-year period of not particularly good times in the field of agriculture, he pointed out with a measure of pride that the farm had made an average net return annually of more than 5 percent on what he described as "a high inventory value." This bespoke good management, and also another factor: that the key men employed on the farm had been with the owner for unusually long periods of time, and that all shared in the financial outcome at the end of each year. This gave them added incentive to do their jobs as well as possible.

Two of these men, and the best known, were George Collister, a Manx lad whom the General found in Painesville and sent out to manage the place, and his son, John Collister, who succeeded to the post upon his father's death in 1933, after 52 years at Juniata. John Collister continued as faithful foreman at the farm throughout the remainder of Dan Casement's lifetime, and, in fact, lived on until 1974. Both the George Collister and John Collister families at Juniata lived in the old sandstone house which was on the place when General Casement acquired it. The Casement family never resided at the farm.

CASEMENT OF JUNIATA

A rare brand of loyalty and appreciation, freely and generously extended in both directions, always existed between Casement and the Collisters. "With all honor and respect, I will say that these men's lives have been far more than incidents in my own," he once wrote, "and, in truth have been a major factor."

As soon as the books were closed on a year's operations Casement would work at his desk late into the evening of each day until the annual statement, inventory and budget were finished and typed. "Today," he wrote on a February 14, "we posted up in the barn a bulletin to the men on the farm summarizing the result of last year's operations, and then gave to each an added check representing a percentage of the wages earned by him during the year identical with the rate of return to me on my personal investment in the property—a practice which has been in vogue on the farm for many years." It was not uncommon for this bonus to amount to $500 to $600 for each of the approximately 15 or 16 persons on the payroll.

His personal relationship and rapport with his employees and their families appears to have been similarly unusual and satisfactory, and on at least one occasion he manifested his appreciation in a manner which struck a particularly happy note. Coming home from a trip west, he wrote of the circumstances of his arrival and the culmination:

"Kansas was still hot and dry but I found the crops on Juniata standing it surprisingly well.

As a Man and as a Stockman

After a few more anxious days we had a gentle rain. To celebrate our good fortune I invited everybody on the farm to go with me to the circus. Men, women and small fry, we tallied 22 head. Ringling Bros. put on a show that fully justified the announcer's resonant statement: 'You have never seen anything like *this* before!' It was certainly a super-circus. As it climaxed in the grand finale with 30 happy, fresh-faced, pretty elephant girls, each wearing a jaunty fez surmounted by a gay cockade (and little else), disporting themselves gaily and gracefully over their massive mounts, I gave vent to my enthusiasm with a stentorian cowboy yell that rang to the apex of the big top."

Contrary to what a good many probably thought, Casement was not wealthy in the sense customarily applied to those who bring city-made fortunes to rural areas for use in developing show places. Rather, it would be more accurate to say that he was "well off," or "moderately well to do." But not rich in the usually understood sense, and normally dependent in his operations upon a line of credit which he carefully tended with a bank in Cleveland, near his Ohio birthplace, and another with the livestock commission company with which he long did business.

Juniata Farm was clearly designed as a utilitarian plant for production of livestock and the feed crops needed by the animals with which it was stocked—the cattle, hogs, sheep and horses. Facilities were commodious but obviously were constructed for practical use rather than for show. "Nothing fancy," the owner said to one visitor, as the sweep of an arm indicated a set of neatly grouped buildings—low, wide and wholly business-like.

The barns, lots and sheds had a capacity of around 600 head, but seldom did Casement have on hand anything like that number, especially in his latter years. In addition to the steers seen at leading shows, others were fed out and sent straight to market. Feed storage was provided by a modern elevator, in conjunction with which there was a grinding plant. The buildings, including comfortable dwellings for the employees, plus a horse barn, were grouped at one end of the farm, with loading pens and a Union Pacific railroad switch named Casement at one edge.

This section of valley land, bounded on the north and east by a right-angle bend on the high side of the Big Blue River, and thus seldom subject to serious flooding, is so level that, as Casement many a time said, "You can see a jackrabbit anywhere on it." The principal crop was corn, key ingredient in his cattle- and hog-feeding program, but he followed a well-balanced rotation which included small grains, hay, and Atlas sorghum for use as silage.

His 2,400 acres northward in the hills were heavily grassed and well-watered pastures. This tract was fenced and cross-fenced exactly as the owner desired in order to provide maximum

As a Man and as a Stockman

utility. The Hereford breeding herd maintained there usually numbered from around 275 to 350 head, and occasionally a few more. After 1912 there was always a purebred unit, at times numbering as many as 100 head, with the remainder consisting of high-grade commercial cows. In the herd's later years, the Colorado Domino and Mill Iron influence, as developed under the guidance of his long-time good friend, Lafayette M. Hughes, was predominant. In addition to the breeding cattle, a good set of steers sometimes could be found on the pastures, and usually a band of Quarter Horse mares and one of the studs.

Dan Casement had a peculiar feeling for these hills and was inexorably drawn to them whenever he felt the personal need for consolation and comfort. At such times he would drive to the pastures, catch and saddle a mount, and ride among the cattle.

"I have seldom tried to analyze the impulse that leads me when troubled in mind," he once wrote, "to seek the solace of these peaceful pastures. It is not, I am sure, the trifling feeling of omnipotence that creeps into my consciousness when I look at all the young things whose lives I have ordained and directed. That feeling would be quickly annihilated by the knowledge that I could no more have stayed the hand that hurled the thunderbolt which recently killed a cow than I could have created the life germ that caused the first tender blade of corn to push its way through

CASEMENT OF JUNIATA

the brown furrow and reach up toward the sun.

"It may be because, up there on the grass, one is more fully aware of the majesty and constancy of Nature's authority and feels that no matter what may come in one's own little life—'God's in his heaven; all's well with the world.'"

The above reference to corn is a reminder that he could wax eloquently philosophical—or perhaps it might be termed romantic—when his thoughts turned to this miracle crop. For example, looking back from the vantage point of 80 full and fruitful years, in 1948 he wrote this:

"I am glad that fate led me to the Cornbelt to make my permanent home. I love the corn. It is to me the most interesting and intricate manifestation of Nature's wonderful wisdom. In a season when its progress is orderly and undisturbed by any cataclysm, the pageant moves so swiftly before your anxious eye from the first weak sprout to the tall, strong, dark-green stalk bearing the heavy golden ears that it seems sheer magic.

"Personally I am inclined to the Chinese philosophy which holds that there is a sensitivity and feeling in all growing things. When my corn field is a mass of tasselled bloom and tender silk, to the plant pathologist, the scientist or biologist it may be an interesting study in the marvelous process of reproduction, but to me it is one grand and glorious love affair."

A few paragraphs later, he continued: "I am

As a Man and as a Stockman

now fitting 110 young Herefords for the coming carlot shows. Daily I see them devouring the beautiful golden product of the acres of blissful mating that was consummated in my corn field under the June moon with my grateful benediction. I imagine that the natural reaction of some of my more practical readers to this observation will be: 'There goes Dan again, wandering about with his head in the clouds!' That I do, thank God! But my boots are always pretty firmly planted in the muck-strewn feedlots of Juniata."

How he did relish the companionship and the challenge found at the stock shows where his Herefords competed, and, most of all, at Denver's National Western. "It may be the tonic of the mountain air," he wrote following his trip to that show in January, 1948, "or it may be the stimulus of the friendly companionship of the mountain men, or it may be because the foundations of my career in the cattle business were laid in Colorado, but, whatever the reason, my heart warms to the Denver show as to no other. My spirit expands with every happy incident."

A year later, upon returning home from the Denver show which marked the 40th anniversary of his debut there as an exhibitor of feeder cattle from the Unaweep, and where four years later he began showing beeves finished in the Juniata lots, he reflected again, not so much this time on the show but mainly in appreciation of

CASEMENT OF JUNIATA

the two places with which his life had been inextricably interwoven—the Unaweep and Juniata. He wrote:

"As I sat sipping my after-dinner glass of port this evening something in its ruddy glow recalled to my memory the little red house nestling at the foot of a towering Colorado cliff. Half a mile away to the south rises a similar granite wall, and in the box-like cleft of the Uncompahgre Plateau lies the beautiful, remote little gem of emerald meadow and meandering stream where, as a youth, just out of college, I learned the rudiments of the business that was to occupy my life.

"I learned my lessons the rough way. High adventure and hilariously undertaken hazards filled my days. Many of my wintry evenings were spent at a table drawn close to the red-bellied, cast-iron stove, writing by the light of a smoky kerosene lamp adoring letters to the beautiful girl who was later to become my wife. My few neighbors were mostly men like myself. There also were some older characters with whom it was both discourteous and indiscrete to evince any curiosity as to their past.

"The winters were often bitter. My memory of them gives me an understanding sympathy for those stockmen who, on occasion, have a desperate struggle with the brutal blizzard. The summers were delightful; game was abundant, even though the Utes, who just preceded me, had taken their full toll of buckskin. In springtime, with the melting of the snow 'on top,' a

As a Man and as a Stockman

waterfall behind the house made a sheer 800-foot leap over the cliff, rushing and roaring its welcome way into our irrigation ditches.

"In many of the later pictures my memory paints, there is a sturdy stone house standing on the bank of a Kansas river—the Big Blue—a house I first saw as a boy of 10 years. It overlooks a square mile of fertile acres, first owned by a Wyandot Indian. This place also is beautiful in a placid, domestic way, although the surrounding Bluestem Hills have not the wild, scenic splendor of the Uncompahgre.

"So interwoven with the fabric of my life are both of these places that, as I sit and dream, my mind flits like a shuttle-cock between the two: The Unaweep and Juniata—the Alpha and Omega of one cattleman's career.

"It's been a good life, and it's a great game. I play it less strenuously than when the riot of youth and stern necessity teamed up to draw me on. I risk fewer blue chips on uncertain hands but, so long as a kindly Fate may allow me to sit in—win, lose or draw—I shall play the game for the sheer love of it."

Following Dan Casement's death on March 7, 1953, Juniata was sold to a feed milling company which used it for experimental research purposes. The only remaining physical evidence in the Manhattan area of the friendly, ruddy-complexioned gentleman—the commanding personality who was large in heart and spirit though modest in stature—who owned this land for almost 64 years is a rather obscure street named

CASEMENT OF JUNIATA

Casement Road. It turns eastward off U. S. Highway 24 slightly more than a mile north of Poyntz Avenue, which is Manhattan's main street, and then bends northward and continues for some three miles to the old Juniata headquarters. The curious may come to look, but Juniata's glory is gone.

If beautiful prose were poetry, Casement would surely qualify as poet laureate of the Bluestem Hills. His classic description of this region, famous across America wherever beef cattle are produced, was first published in his "Random Recollections," and afterwards widely quoted.

"It is hard to describe the Bluestem Hills to anyone who has never known and loved them," he said. "They have none of the rich, flaming beauty of the forested New England mountains; none of the austere majesty of the Rockies. Rather they are Mother Nature's round, undulating breasts, soft and warm in the sunshine, restfully inviting and rich in the promise of nurture."

No one who has not lived in close communion with nature, or who did not have an extraordinary sensitivity to beauty, could have written those words, or these which followed:

"There is always a breeze blowing there—cooling, refreshing, and strong enough to blow the cobwebs from a tired man's brain. There are sumac bushes and buckbrush to be found on top, but trees grow only in the dividing ravines, where little runs are fed by springs

As a Man and as a Stockman

from the hills, and there the cattle find shade and water. There they spend much time in the hot summer days, but when the blazing Kansas sun drops down out of the brazen sky, they usually drift up onto the divide to enjoy in the open the cooling breath of the evening breeze.

"Often a stranger might think the pastures vacant, but they are teeming with life if you have eyes to see and ears to hear. You may catch a glimpse of the slinking figure of a coyote. . . . You may hear the whistle of an upland plover; see an occasional prairie chicken or covey of quail; or you may watch the flight of a great blue heron as she wings her way over the hills . . .

"And, too, there is the seasonal procession of wild flowers: evening primroses, Canterbury bells, the dainty pink ball of perfume we call the sensitive rose, indigo flowers, the brilliant Indian paint brush—and dozens of others. And in the fall, of course, the goldenrod, the wild aster and the purple Kansas gayfeather."

Over a period of several years during the 1920's he was a frequent contributor to *The Breeder's Gazette*, sometimes through a personal column headed "A Stock-Farmer's Diary," and at other times under appropriate subject headlines. Many of these contributions make fascinating reading because of the indication they furnish of the breadth of Casement's interests and frequently the depth of his perception. Some of them also provide evidence of the long hours devoted to farm tasks which enabled him, too, to maintain his close relationship with nature; for example, of a late January day, he wrote:

"Robins made their first appearance at the farm this morning. Tuneful they were, as if inspired by the bright, mild weather which seems to give assurance that the back of winter is broken." Then he proceeded to relate details of a busy day at the farm immunizing fall pigs, dehorning calves from New Mexico's Crosselle Ranch, selling a boar, bringing in a band of 46 horses from the stalkfield, selling a couple of Quarter Horse fillies to a visitor from Illinois, and trading two lambs for a ram and a lamb with the Scottish shepherd at the college. After which, he wrote that he worked at his desk until the middle of the night on the annual farm statement and inventory, "perplexed and harried by multitudes of figures which refuse to balance. As I write at midnight," he concluded, "the sky is starlit and the temperature well above the freezing point."

Incidentally, as he struggled with those figures, it may be significant to indicate that this instinct for precision seemed to have been inborn. Even as a boy at Princeton he maintained his personal financial accounts meticulously. His daily account book, still in existence, lists his daily expenses right down to 5 cents for a newspaper, 10 cents for carfare, 25 cents to a porter and 10 cents for a shoe shine. And at the end of each week his books balanced out right to the penny.

As a Man and as a Stockman

So it was no happenstance that in his mature years his Juniata Farm accounts were maintained with the same exactness, even if he did on occasion have to "burn the midnight oil" to make them come out right.

He was uncommonly aware of the moods and beauties of nature and of the whimsical procession of the seasons which, as he once wrote, "moves on, calm and inexorable." Following a 1926 trip to southern Arizona, he wrote of the area around Globe: "Everywhere the slopes were brilliant with poppy blooms, bright orange in color. Birds of strange sorts abounded, and Gamel's partridge, the common crested quail of this country, called loudly on all sides." About that same time, after looking at calves with a companion on New Mexico's Crosselle range, he wrote that while "homing under the stars and beneath a brilliant moon, our talk turned to blizzards we had known."

On a wintry Kansas night, as the years closed in, his thoughts still turned toward the unplumbed mysteries and unexplained wonders that intrigued him. He wrote:

"Sometime, not too distant, I shall wake up in the night to hear, far up in the sky, that hoarse, haunting cry of wild geese in flight, and I shall feel the same thrill of excitement with which it always fills me. It is not only that it is a sure harbinger of spring; it is the wonder of it all. Whence have they come? Whither do they go in that strong-winged, orderly flight that carries them over so many miles, and will carry

CASEMENT OF JUNIATA

generations of them who cannot know the way from any experience of their own to the same annual breeding ground? And I shall fall asleep again marveling at what Nature hath wrought!

"At the first hint of spring I shall be up and out catching the smell of the good earth as the furrows are turned; seeing the hills change from brown to verdant green; awaiting eagerly the rebirth of all around me with anxious eyes. It is always a new beginning—and a call to head, heart and hand.

"A farmer's life monotonous? It is the fullest measure of changeful living. A farmer needs no subsidy except the will to work, no bounty save to enjoy as he may the fruits of his own labor and the abundance of a generous Mother Earth."

As he wrote of the passing scene, first for the *Gazette* and later for the *Hereford Journal*, spring-time and fall seemed most to furnish inspiration. "Cold this morning," he wrote of a winter day in 1925, "but with a sun which soon warmed the air, and later the day held a faint suggestion of April. Cardinal grosbeaks called clearly from the thickets."

Twenty-three years later, he wrote: "April is a notoriously whimsical jade. So, for that matter, is Kansas. Together they put on a great show. This April morning there was a sharp nip in the air and the warmth from a light fire on the hearth was quite welcome as I sat before it smoking my after-breakfast cigar and reading the morning paper. This leisurely custom is a slight concession to the encroaching years. Per-

As a Man and as a Stockman

haps my sense of duty grows more elastic as my steps grow less so, but I no longer consume my news and coffee in alternate gulps. The paper today was so filled with battle, murder and sudden death, political tirades, world economic distress and revolution that I found myself wondering how this sound and fury might culminate . . .

"It was a depressing thought, even for a man who likes a little excitement, and I stepped out of the house in a somewhat melancholy state of mind. Stepped out into a world bright with April sunshine and a blaze of gaiety and color from the tulips massed before the house! There couldn't be much wrong in a world where such resurrection is annual, I thought in reflection."

The inspiration which he seemed always to feel with the annual coming of spring was highlighted again in these words written in the latter 1940's after a ride through his Bluestem pastures:

"Always with the advent of spring, a consciousness of the miraculous awakening and renewal of life in his world has brought to man an instinctive feeling of uplift and encouragement—the elation and satisfaction that greet the fulfillment of a promise long deferred. Since the beginning of time man has celebrated this season with Easter festivals and ceremonials expressive of his joy and his gratitude to whatever gods he may worship for achieving this annual resurrection. The thoughtful and reflective observer of this miracle finds his spirit renewed and strengthened in a measure that varies with his near and remote contact with the natural processes that attend its consummation. The farmer and the husbandsman are favored in this regard to an extent the urbanite can never know, because they are direct and active participants in this process of creation and fruition."

This matter of fruition, the gathering in of the harvest, was on Casement's mind as he wrote in the fall of that same year:

"It is not strange that Kansas stockmen and farmers commonly regard October as the best month of the year. It is their month of realization and fulfillment. What seasonal reward Nature and his own industry and intelligence have brought to the man on the soil is now actually achieved or can be pretty accurately appraised. October weather in Kansas usually is pleasant. The landscape wears its gaudiest garments. It's a good time and place wherein fully to enjoy the supreme delight of just being alive . . . By and large, Nature has been beneficent and kind to us throughout the season now closing and we face winter with more than the accustomed sense of security and well-being."

In his own distinctive way, Dan Casement was a man of great faith. To him, as Dr. James C. Carey of Kansas State University once pointed out, there was a sort of sacredness in every living thing, from the "highest mammal, man," to the tiniest wild flower. "He was quite unusual," Dr. Carey went on, "in the fact that in a very unconventional way he was a very religious man.

As a Man and as a Stockman

He believed completely that 'there were God-given laws which men could not violate with impunity.'"

He was eloquent and expressive about the miracle of new life. A spring-time visitor once accompanied him to the farm and together they observed the wobbly baby calves, the leggy new colts, the fleecy little lambs and dozens of inquisitive pigs, all distressing their nervous mothers with their unrestrained exuberance. The visitor commented on the marvel of it all.

"You're right," Casement said, "it's like the resurrection. In spring-time," he added softly, "I'm the most devout son-of-a-bitch you ever saw."

Basic elements in his personal philosophy also were apparent in his *Gazette* writings, and on these he expressed himself with a vigor, and sometimes a sense of indignation, that few could match. Although a veteran of the first World War, for which he had volunteered although far past military age, he adamantly opposed the 1924 bill which granted a bonus to those who served in it, and decided to run in his district for the U. S. House of Representatives against the incumbent who had voted to override President Coolidge's veto of the bill. His sense of patriotism showed in this *Gazette* comment:

"I opposed the bonus on principle. My conception of the matter was that when we, who had acceptable bodies, offered them and our lives to the country, we simply fulfilled a pressing obligation and accepted a real privilege—that of offering the supreme sacrifice in a supremely righteous cause. The only ennobling circumstance associated with war, as it seems to me, is its imperative call to men to make this sacrifice. Surely, on a belief in the efficacy and nobility of such sacrifice, Christianity and the whole structure of the highest civilization is based. To those, then, hampered by no restraint of age, infirmity or other circumstance, to whom the privilege of offering this sacrifice was granted, there accrued a positive advantage over their less fortunate fellows. To attempt to compensate them materially for an offer transcending any material reward, as I conceive it, not only robs them completely of this advantage but also insults the only rational motives on which their offer could have been made. To weigh the worth of a soldier's sacrificial offering in terms of cash or life insurance seems to me to savor of sacrilege."

As a sportsman as well as a farmer-stockman, he drew this intriguing parallel as he "hunted" for voters: "At picnics and other functions of a non-political nature they may occasionally be surprised, and perhaps some of the less alert and less wary birds may be winged or even brought to bag, as the covey flushes, but usually the sport reverts to a search for singles, lying close to a dense cover of secretiveness, sometimes thorny with resentment."

This characterization of Casement which many of his old friends will agree might have

As a Man and as a Stockman

been equally applicable at other times and places appeared in *The Manhattan Tribune* at the height of that congressional campaign:

"He enjoys life hugely and he enjoys people. He will be darn glad to talk to and meet you regardless of your support for him. Chances are that he will find something a lot better than votes to talk about. When Dan comes into your vicinity you will think the circus has just got into town. The band is beginning to play, the small boys are giving three ringing cheers, and everyone is having a bully good time. He wears red neckties and any loud clothes he happens to grab first when he wakes up in the morning. Of course the committee may curb and curry him a little for the campaign but Dan is pretty sure to interest you unless you are an awful sour old fossil.

"If Dan is in the crowd there will be some noise. Dan won't make it all. The whole bunch peps up when he is around. Of course he knows the conventions, but, Oh Boy, he don't (sic) let them cast a gloom over the occasion and spoil a good time. So, if you hear any hilarious noise in your part of the country it is probably Dan meeting the folks and having a good time. Don't miss seeing him. He is like a fresh breeze on a sultry day."

But as might have been expected in a time of economic stress, the majority voted for cash, so he lost to the incumbent in this, his only bid for state or national office. It was congenitally impossible, or virtually so, for him to straddle an issue, and once his decision was made it was instinctive for him to go all-out in its behalf. This all but foreordained that he would not be a success as a politician. He did, however, render national service when he was appointed in 1926 by Secretary of Agriculture W. M. Jardine to review an appraisal of the grazing value of the National Forests. After a detailed study, he recommended that the fees charged for grazing be related to the prices of livestock, a plan that shortly went into effect and long was accepted as a fair practice. He also responded on occasion to the call for community service, as for example, when he was named county chairman of the American Red Cross in the 1930's, but he abstained from ever again seeking elective office.

But in office or out, no one ever needed doubt where he stood on issues of the times. His concern for the rights, opportunities and responsibilities of the citizen never wavered. He was never reluctant to enter the fray on matters involving politics and public affairs, rights and responsibilities, liberty and freedom.

He vigorously condemned the Ku Klux Klan as it made a strong though unsuccessful bid to establish a foothold in the state. He incited his long-time personal friend, William Allen White, famous editor of the equally famous *Emporia Gazette*, to spearhead the effort which led to the defeat of Dr. John R. Brinkley when the so-called "goat-gland specialist" sought to become governor of Kansas. But it was his outrage over the federal government's encroach-

As a Man and as a Stockman

ment upon personal freedom and individual liberty, as perhaps first expressed in an extended exchange of letters with the Emporia editor during a 16-year span, that brought him to broadest national attention.

He gained renown as "the great dissenter" through his opposition to the farm-support programs espoused by Agriculture Department Secretary Henry A. Wallace during the Franklin D. Roosevelt administrations in Washington. He literally bombarded federal officials with protests against the ethics of the Agricultural Adjustment Act, its program and its policies. He spoke, wrote and debated in such broadcast forums as the Town Meeting of the Air with such New Deal advocates as Thurman Arnold against what he believed to be a proposition that was wholly wrong. Not only did he denounce the AAA and its regulations, wheat quotas and the like, terming the subsidies offered as "contemptible bribes," but he refused absolutely to accept the benefit payments that could have been his under the law had he chosen to place profit over principle. At the height of the controversy he had referred to the Washington program as a plan "that robs collective Peter to pay selective Paul," and added, "I don't know just what commandment it breaks, but, by God, it's stealing."

Casement grew up in the cattle business with western ranchmen of the old school, and his own personal instincts, along with his contacts with them from the 1890's on, encouraged codes, creeds and customs in him which made him oppose the restraints of increasing governmental activities which he found rushing pell-mell in his direction in the 1930's. As the era of the old cowmen and their unique way of life were confronted by a strange new philosophy which was anathema to the principles he had pursued as boy and man, he became a vigorous spokesman in behalf of the rugged individualism of the old range days as a bulwark against governmental interference and additional restrictions which he firmly believed would lead the America which he loved down the primrose path to disaster. In his view there was no middle ground. The two philosophies were entirely incompatible and thus he came "to play the role of the rebel in the days of the New Deal," as Dr. Carey once expressed it in a Kansas City speech.

At the height of this activity, in the middle to latter 1930's, major articles written by him appeared in such national publications as the *Saturday Evening Post, Nation's Business, Fortune, Review of Reviews, Country Home* and others, as well as frequently in the agricultural and livestock press. His fine flair for colorful and vigorous expression and his reputation as an erudite and articulate spokesman for ranchers and farmers in resisting government encroachment upon their operations led to speaking engagements before the Executives' Club of Chicago, and similar gatherings of business and community leaders in such cities as Cleveland, New York, Kansas City and numerous others. He served as president of

As a Man and as a Stockman

the Farmers' Independent Council, a national organization which opposed the government's restrictive and subsidy policies.

"Efficient farmers should not have to meet the unfair competition of high-cost producers sustained by government subsidies," he said in a Chicago speech in 1935 in behalf of this organization. The following year he assailed the government's soil-conservation legislation, adopted as a substitute for the AAA, as a device to enable the Roosevelt administration, as he put it, to give money to farmers to win their political support.

"The administration," he said, "must be bankrupt in civic morals, common sense, and patriotism when its only prescription for every social and economic infirmity is public money in ever-increasing doses," reported the *Chicago Tribune*.

The essence of his own program, as reiterated in what he said and wrote on the matter, was for farmers to produce all they could, practice thrift and economy, tend to their own business, take agriculture out of politics, and stop looking to the government for help.

His credo regarding the effect of Roosevelt's New Deal was fairly well summed up in this paragraph from one of his letters to William Allen White:

"Our disregard of natural laws has accomplished this debacle. Legislating artificial aid to organized labor through fear of its political power helped to put the price of tractors above the reach of the prudent farmer. Now the farmer goes on the prod and government hands him subsidies as a sop to keep him quiet but dares not attack his problem man-fashion by undoing the wrongs it has done him in boosting the cost of necessities by stupidly trying to substitute human for natural law in urban industry."

This was a point he tried desperately to convey. In a philosophical article written in 1938, he emphasized that determination of the prices a farmer-stockman receives for his products is governed by natural forces. "The volume of his production is determined by nature. His home price is primarily fixed by the relation which that variable volume bears to a variable demand. Both of these factors are beyond his control and cannot be altered by any artificial device.

"But, on the other hand," he continued, "prices of the goods and services the farmer buys from urban industry *are* subject to human control. It seems clear then an equitable relation between agricultural and industrial prices *can be established only by adjustments in the latter.*"

The thrust of his expression was that industrial wages should be tied to farm prices, rather than vice versa. "No injustice would be involved," he asserted, "nor would the real wage and standard of living of the workman in industry be lowered. For, should nature be lavish and food and fabric consequently low in price, a comparatively small money wage would then buy as large a quantity of the raw necessities of life as a higher wage would buy under contrary conditions."

As a Man and as a Stockman

Acknowledging that wages and prices were being largely fashioned by human will and human avarice, he proposed, as if in answer to those who may have felt that he was obsessed with negativism, that "if the problem of equity as between the farm and the factory were calmly and realistically approached from the east end, so to speak, it would stand a better chance of satisfactory solution than is offered by the method and direction of approach heretofore employed."

From the detached seclusion of his feedlot and out of such wisdom as his beasts had taught him, he proceeded in this 1938 comment in *The Kansas Magazine* to reiterate this counsel:

"The concise principles on which all right-minded citizens, whether sane liberals or sincere conservatives, should unite in organized opposition to Roosevelt's ruinous program must be in complete harmony with established truths and absolutely concordant with the fundamental laws of Nature. Otherwise we shall have only a flimsy footing on which to stand and fight for our own and our country's salvation."

In a 1941 interview with the editor of *The Cleveland News* he characteristically urged his countrymen "to shuck off all of the soft-bellied human folly we have piled up," and 10 years later in a speech at the 30th annual meeting of the Panhandle Plains Historical Society at Canyon, Texas, when nearing his 83rd birthday, he showed that his resolve was undiminished by pointing the finger of scorn at those who would seek "the satisfaction of their belly needs at the

CASEMENT OF JUNIATA

sacrifice of their morals and their manhood." And he voiced once again, and possibly for the last time in public, his life-long thesis that "man's impulse for freedom . . . is an instinctive adjunct of his very nature."

Although his zeal in pursuit of this ideal never lessened, except as preordained by the passage of time, one of the few disappointments of his life must have been his failure to stem the tide "of those who profess to discern error and injustice in Nature's plan and virtue in dogmas which Nature plainly abhors." But he continued steadfast in his own personal belief.

Nor was it in Casement to countenance sham, pretense or hypocrisy. Even less could he "stand still" for any man he felt was playing loose with the truth, but he could do it in a gentlemanly manner, for, as one reporter put it, he possessed an ingrained polish and gentility that his many years of living in the wide open spaces dimmed not one whit.

Bill Colvin, editor of *The Manhattan Mercury*, who went to Manhattan as a young man to work on the newspaper a few years before Casement's death, once recalled a characteristic comment which Casement occasionally applied.

"One of Dan's favorite descriptive expressions," Colvin wrote, "when he was really teed off at an individual, which wasn't too often but when it was it was with colorful vehemence, was to exclaim, 'He's a prismatic s. o. b.' When the uninitiated would ask Dan

As a Man and as a Stockman

what a prismatic s. o. b. was, he would elaborate by observing that the person in question was an s. o. b. from any angle you looked at him. "How's that," Colvin concluded, "for descriptive language?"

Another facet of this unusual man was displayed about the same time by a few humorous lines he wrote following a trip to a Kansas City hospital. "When I was borne into surgery and saw a charming, feminine anaesthetist awaiting me, needle in hand, I was inspired to burst into song, paraphrasing a lyric from 'The Mikado':

'When a man's afraid a beautiful maid's
A pleasing sight to view,
And, Oh! I'm glad that moment sad
Was soothed by the sight of you.'

"Alas! on my second trip to the operating room," Casement continued light-heartedly, "this lovely lady did not officiate. A less expert and far less attractive mere male drilled in my spinal region like a woodpecker in vain search for the vital spot."

He was the author of the widely distributed "Holding to Freedom" resolution presented at conventions of the American National Cattlemen's Association, the Kansas Livestock Association and others in the late 1940s as a protest against the stifling of individual initiative and free enterprise, as he viewed them, and probably deserved as much credit as anyone, if not more, for the fact that the cattle industry was the only significant segment of agriculture that did not eventually come under restrictive government controls.

In presenting this resolution in the deeply resonant voice which he could use with the skill of a fine organist, he called upon patriotic Americans to denounce "the fallacious policies that are beguiling our country into socialism" and to "work energetically and courageously for the reestablishment of free and competitive enterprise in the restoration of the Republic." The policies he condemned with such vigor, he said in characteristic conclusion, were "in impious contempt of Nature's eternal laws."

Casement usually was more likely to be found in the corridors or halls outside the meeting rooms than in attendance at the formal sessions when he went to the various association conventions. In once relating how supremely happy he was upon arriving at a convention city to find so many "dear people who share my outlook and my enthusiasm," he went on to say:

"It has always seemed to me that actually the main and most logical purpose of these conventions of cattlemen is to bring together a group of the most admirable human beings, not so much to discuss the so-called problems of their industry as to cultivate friendships and to strengthen the bonds of respect and admiration for the pattern into which mutual pursuits and experiences have molded their characters."

He may have been considered something of a maverick upon occasion, for when he disagreed even with the majority he did not hesitate to

As a Man and as a Stockman

say so. But they loved him just the same. For example, following one of the national livestock conventions he attended, he wrote:

"In the circumstances, it was worse than a waste of time to pass at least 28 of the 30 resolutions which the convention finally adopted. Expediency so dominates the national political picture and our whole economy is obviously so near collapse that it seemed to me absurd to pass a string of resolutions which almost no one would read and which could not, in a crisis so acute, make any impression on or receive any consideration from those to whom they were addressed."

He had an innate capacity for friendship, attributable in part perhaps to the influence of his father, the fabled General Jack, of whom he once wrote: "Until I was 41 I had the great good fortune to enjoy my father's close companionship. He was the most delightful comrade I have ever known. His humor was gay and irrepressible; his generosity unbounded; his courage unfailing, and his common sense profound." With this inherent instinct which was lovingly cultivated, it would have been surprising if the son had not had so many friends—those drawn to a cultured gentleman of gracious charm who was equally at ease in the drawing room or on the roundup. And who made his friends, from whatever walk of life, equally at ease.

"I have learned what a precious gift is life itself," he once said. "I am everlastingly grateful to my parents," he added, "for bestowing it upon me," and then concluded: "The longer I live the more delightful I find life and the more eagerly I aspire to keep on living on the face of this gorgeous earth."

In addition to regular attendance at the American National's annual winter meetings, he would be in Texas almost as often for the spring convention of the Texas and Southwestern Cattle Raisers Association. There was some indication that he was drawn there, in part at least, by the advance look the trip would provide at approaching spring-time. In describing the passing scene from a railroad car window as he once journeyed to Fort Worth, he observed: "Peach trees were blooming in the door yards of ranch houses and wild plum thickets were whitening in the sandy draws. New leaves on the trees showed faintly green along the creeks." And, of course, as a stockman, he noted the condition of the cattle and pastures seen along the way.

Naturally the Kansas Livestock Association meetings drew him, as did the conventions of the Colorado cattlemen, he having been a vice-president of the Colorado group when he still had the Unaweep range on the Western Slope. In fact, when the Colorado association honored Theodore Roosevelt on one occasion he was the dinner companion of the former President, whom he had previously met in the White House through the good offices of Rock Channing, a

As a Man and as a Stockman

Princeton classmate. He also found joy in the meetings of the New Mexico Cattle Growers Association, from many of whose members he had bought feeder cattle, and delighted there in the companionship of such old friends as O'Donel, Springer, Mitchell, Captain Mossman and a great many others.

In reflections upon many contacts during a period of more than 20 years with this most vigorous and colorful personality, it is in order to draw upon a personal recollection to emphasize again the outgoing nature and capacity for friendship of this uncommon man. These traits were demonstrated to this writer as never before during a 1948 trip to the American National's convention in Boise, Idaho. Casement later wrote for the *American Hereford Journal* of the trip and some of the pleasures deriving from it in these words:

"While patiently awaiting the miracle of the glad awakening that annually attends the advent of April in Kansas when

> '*The grey March winds are folded safely by,*
> *And rising clouds of blackbirds make*
> *Dark silhouettes against the morning sky*'

we have contrived to brighten the dull days of winter by making a joyous excursion. On January 11 I took the train to Boise to attend the 51st annual meeting of the American National. My pleasure began when, on leaving Denver, I discovered that a fellow passenger, bound for the same destination, was the genial editor of this magazine. Another ray of sunshine brightened our day when, at Cheyenne, we met up with our good friends, the Haywards of Cimarron, New Mexico, and, when Bill Sidley of the Silver Spur boarded the train, with some others, at Rawlins, my cup of pleasure was full to the brim.

"When, on landing at Boise, I found Frank and Mary Boice, Dub and Beulah Evans, Huling and Hester Means, and dozens of other dear people who share my outlook and enthusiasm, I was supremely happy and rejoiced that I had come."

After witnessing at close range this demonstration of the breadth and warmth of his friendships, one could only be amazed that one person could know so many from such a wide area so well and could so delight in seeing them again. It was an unforgettable experience in the realm of friendship.

"I lift my glass to friendship, than which there is no finer wine," he once said, and, of the men with whom he sipped at such stockmen's gatherings, he commented: "They bring with them the fresh air of their high home ranges; the salty idiom of the cow camp; the moral strength, courage, common sense and integrity of the pioneers who built America . . ."

It was the impression left on the youthful mind of an Iowa farm boy by the eloquence of Casement's words as published in the 1920's in *The Breeder's Gazette* which led in the 1940's to the writing of his "Random Recollections" for the *American Hereford Journal*, and to a

As a Man and as a Stockman

subsequent series on contemporary matters which appeared under the heading of "Juniata Jottings." The idea for the first series was proposed to him during a visit in the lobby of Chicago's Stock Yard Inn during the 1944 International stock show by the one-time "farm boy" who had just become editor of the *Journal*. Casement's reaction was no more than lukewarm and he committed himself at that time to do no more than think about the matter. In a follow-up letter to him a complimentary reference was made to a little book about himself and his family that had been published about a year earlier for members of his family and a very, very few others—*The Abbreviated Autobiography of a Joyous Pagan*. And, of course, the invitation to write was renewed.

His reply, succinct and pithy, and reflecting his charming sense of modesty, went like this:

"I'm delighted that you liked the 'little book.' And your suggestion that I try my hand at a series of reminiscent articles rather intrigues me, though it would be a hell of a job and I seriously doubt my ability to do it decently.

"Suppose we leave it up in the air and, if I find the time, perhaps I'll try it out. It occurs to me that my recollections of W. J. Tod might make a readable story. Maybe I'll take a whack at it some day.

"But anyhow, I'm immensely pleased by your overappraisal of my literary ability and your approbation certainly expands my ego."

Within a month a letter came advising that an article was under way on the Scottish-born Mr. Tod, whose ranching activities in the Southwest and cattle-feeding activities at Maple Hill, Kansas, had made him one of the most prominent stockmen of his time. But Casement still had some misgivings, unfounded though they were, as indicated by the paragraph that follows:

"Writing is always a bitch of a tough task for me, but I loved Mr. Tod and, if I have the luck to get the right scald on her, she'll be, I think, no slouch."

A few days later the best article ever written about Mr. Tod arrived, accompanied by a letter in which Casement said:

"To be quite frank—and utterly shameless —about it, I think I got a right good scald on her. I hope you'll agree. Anyhow, I worked like hell. Writing makes me sweat."

The series became an immediate hit and attracted a wide response, which was highly gratifying to him and to the editor as well. Casement continued to write more or less regularly for the *Journal* for the remainder of his life, the 30th and final installment of the "Juniata Jottings" series being received only three days before his death and appearing in the same issue in which his passing was reported.

This prophetic paragraph was included near the opening of that final installment: "This will

As a Man and as a Stockman

be our last contribution to the *Hereford Journal*, because Nature is having her inexorable way with my body after 84½ years of joyous living in close association with men of the range, ranch and feedlot. These are the people I like. This is the life I have loved." Shortly thereafter came the telegram from his son Jack stating that the end had come.

The Casement letters, always typed on letterheads printed entirely in red ink, were expressive and illuminating, and even yet some are a delight to re-read. For example, in one advising that an article which had been under consideration would be written next "if God spares us," he appended this postscript:

"Nature is not being very kind to us this season. High winds and variable temperatures, with a sad dearth of moisture, make our position very vulnerable. But I figure that is God A'mighty's business and I can rationally ask only for composure to endure."

This philosophical acceptance of a situation about which he could do nothing was far more expressive than had he merely stated that "it is dry out here."

Nor did he hesitate to take the editor to task when he deemed such to be in order. As one who regarded his animals as individuals, and each one almost as a personal friend, he once wrote to protest a change that had been made in one of his manuscripts:

"I have a personal animosity," he said, "against making reference to cattle and horses

CASEMENT OF JUNIATA

impersonally. Always I use the personal pronoun *who* instead of *which*. I notice you made this change in my opening paragraph. Would it play hell if you humored me in this regard in the future?"

In the big house on Humboldt Street, a block north of Manhattan's principal business street, which was the family home after 1926, were inscribed pictures, letters and books from many of the great and near-great who had enjoyed Casement hospitality there. A time-worn letter from Theodore Roosevelt, whom Casement greatly admired as President of the United States, and perhaps as something of a kindred spirit, regretted that he would not be able to make an anticipated visit to the Casement home because of the press of other matters. These mementos came from many places and reflected his friendship with such colorful personalities as cowboy artist Charles M. Russell and the inimitable Will Rogers, as well as with financiers, writers, military figures, politicians and others, including, of course, many prominent stockmen.

This house, with its drawing room, music room, sun porches, garden area with fountain and pergola, plus other accoutrements of gracious living and entertaining, continued to be his home after Mrs. Casement's death in 1942. It was in the large living room there, surrounded by members of his family and some close friends, that his funeral was conducted on a blustery

As a Man and as a Stockman

March morning in 1953 before the final trip was begun back to Ohio and a resting place at Painesville.

The Casement house, through many years of its owner's lifetime, was an occasional gathering place for the local Play Reading Club which now and then brought some of the culturally inclined people of the town and the university to his home where, in his well-worn tuxedo which he could wear with a casual flair, he joined them in renewing an appreciation of the classics. He had a keen liking for good writing, as shown by his well-filled bookcases, and a profound respect for the works of Kipling, Lord Byron, Alexander Pope, Kingsley and others, including such skillful interpreters of the Old West as Eugene Manlove Rhodes.

One long-time Casement friend, indeed, wrote following his death that to her he exemplified the hero of Rhodes' famous poem, "The Hired Man on Horseback," and that never again would she read without thinking of him such lines from it as

*"Doggerel upon his lips and valor
   in his heart,
Not to flinch and not to fail, not
   to shirk his part;*

or these:

*"Hat tip-tilted and his head held high,
Brave spurs jingling as he passes
   by . . ."*

Some of the works of the various writers which he especially liked, he could quote at length, and his own articles and speeches were liberally interspersed with appropriate quotations. He was, indeed, a literate as well as an articulate man.

Casement was remarkably flexible and adaptable, which are not always characteristics of those of advanced years. A couple of years after the beginning of "Random Recollections" in the *American Hereford Journal,* he invited the editor and his family to come out from Kansas City for a Sunday picnic. "We will have a lunch prepared and all go to the hills to eat it," he had written. The family, and especially the two young sons, looked forward with great relish to this outing in the Bluestem Hills, and all, of course, were dismayed when a heavy rainstorm broke shortly before Manhattan was reached.

But was the day ruined? Far from it! One of the most memorable of picnics was held right in the Casement home, with the warmth and hospitality of the host making everyone virtually forget that this hadn't been the plan from the outset. What a pity that the old house finally yielded to the onslaught of so-called progress when the property was sold for commercial use and the house moved from it late in 1973!

Many honors came to Dan Casement, and he wore them easily. He served at various times as an officer and/or director of numerous state and national stockmen's organizations of which he was a member. He was named posthumously in

As a Man and as a Stockman

January, 1958, to the National Cowboy Hall of Fame at Oklahoma City, where homage is paid to great men of the West, one of the first two Kansans to be so honored. He was feted by the American Hereford Association in Kansas City for his dedication to the advancement of American animal husbandry. He was a life-time honorary member of the American National Cattlemen's Association, which he served for many years as a member of its executive committee.

He was guest of honor at a luncheon given by the rural magazine, *Country Gentleman,* in Chicago during the 1952 International stock show, during the course of which a speaker said: "If you say it with affection, you may say that Dan Casement is a crusty individualist who believes that American agriculture was built in the fine tradition of self-reliance. We say it with affection. He has the cowman's belligerent desire to be let alone in running his business, yet he is the kindliest of friend to countless thousands of America's cattlemen."

When Jess C. Andrew, then president of the International show, observed that Casement had done "as much as any man in America to advance the cattle industry," the Kansan characteristically disclaimed the right to any such credit, but added with a twinkle that he "loved the sentiment which made you say it."

The honor he prized most of all had come a few years earlier, when his portrait was unveiled in 1939 in the Saddle and Sirloin Club's celebrated gallery of immortal figures associated

CASEMENT OF JUNIATA

with the livestock industry. The citation on that memorable occasion, staged in a banquet hall of this famous Chicago club, referred to him as "Stockman, Farmer, Patriot, Exponent of the Pioneer Self-Reliance, Personal Freedom and Individual Independence That Make the American Tradition."

Writing some years later of that event, he said: "I well remember the evening they 'hung' me in the Saddle and Sirloin Club, and how heart-warming it was on that occasion to be surrounded by all of my family and to look into the faces of many of my dearest friends. . . . It was particularly gratifying to me that my portrait was placed in the group representing men of the old range days whom I honored and loved, among them M. C. Campbell, W. J. Tod and Murdo Mackenzie. Who would not feel honored to meet posterity in such a goodly company?"

As an old friend once expressed it, "The homing instincts that hold men true to great ideals and strong tradition" drew Dan Casement back to Princeton University a great many times through the years. He was there for the last time in June of 1951 when he sat in the sun on the bleachers to witness Yale's defeat in the traditional baseball game, a fact which delighted him since as a student he had been president of the Princeton Baseball Association. This, he once said, meant more to him then than having a seat in the U. S. Senate would have later.

As a Man and as a Stockman

As a graduate of 61 years standing, he later wrote that he "still found zest in the many time-honored activities of the week-end," but even more gratifying, he added, it was to find that after so many years of active, happy living he could "still thrill to the subtle spirit of the great day." In thus making one more excursion into the "full-flowing stream of life," as he expressed it, he was enabled again "to feel the strength of its current in the veins of youngsters whose affection is the supreme reward of my many years."

"As youth recedes," he wrote, "the cares, anxieties and responsibilities drop one by one from our shoulders, leaving us a new freedom of action and expression. I am reaping the emoluments of age. In the society of these younger men I find increasing satisfaction and happiness. They laugh hilariously at my twice-told tales. They listen with respect and apparent admiration which is, I trust, not entirely feigned, when, in my anxiety for the future of this land we love and want to preserve, I expound and expand the thesis that man is a moral animal and freedom is his birthright. They hail me as 'Uncle Dan,' and it warms my heart. I have reached the age where I may take without embarrassment what they so freely offer, with only the humble hope that I may deserve it as fully as I return it. Lord, but it is good to be alive, still active, half Celt, and wholly uninhibited."

On his return home after this last jaunt eastward, which also included a stop at the old

CASEMENT OF JUNIATA

family home, Jennings Place, at Painesville, a visit in Cleveland with "Tot" Otis, the "bosom friend" of his youth and manhood, and a sentimental first-time excursion to Geneva, New York, where his father was born, Dan Casement wrote this touching paragraph:

"When, after 3,000 miles of happy wandering, I drew up to be greeted by the lights of home shining forth their welcome, contentment enveloped me like a garment. As they drew me within the shelter of my own roof tree, my heart swelled with gratitude as I thought life has been good to that man who finds light, comfort, peace and affection awaiting at the end of a long journey."

In summarizing the life of this remarkable man, Dr. Arthur D. Weber, long-time Dean of Agriculture at Kansas State University and later its Vice-President Emeritus, who probably knew Dan Casement at close range over as long a period as anyone, responded as follows in late 1973 when asked to share his recollections:

"I have many impressions of Mr. Casement gained in associations with him while I was a student, a teacher and research worker, an academic administrator, and as the steer judge for 11 consecutive years at the International Live Stock Exposition, most of that time under his watchful eye as a member of the board of directors of that great institution.

"My impressions of him that have survived with the greatest impact relate to his innate

As a Man and as a Stockman

honesty and integrity. All that he said or did reflected his honest convictions and were expressed with great gusto in his own inimitable style. He was profane, yet his profanity was tempered with cultural overtones that made it acceptable even to those obsessed by their own sanctity. He was a cultivated intellectual, yet some of his oldest, most trusted friends were common folk whose basic values and beliefs he could share and articulate.

"Not everyone agreed with his political philosophy. Yet he was universally respected and admired for facing issues squarely and honestly and with forthright consistency. He was truly a great American. His patriotism illuminated all he did and made him one of America's most unique stockmen."

JUNIATA

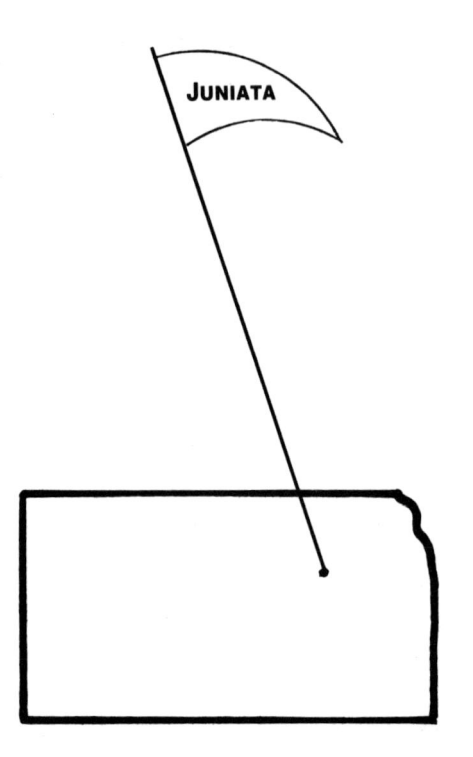

DON ORNDUFF has an appreciation of the wide, open spaces and the people who populate them dating from boyhood days on an Iowa farm. Opportunities for expanding his horizons multiplied after he became a member of the staff of the American Hereford Journal in 1930, a connection that was continued until 1970 when he retired after a quarter-century as editor of that magazine whose sphere is this most populous breed of beef cattle and the men who breed and feed them.

It was in this context that his acquaintance and friendship with Dan Casement evolved over a period of over 20 years. He witnessed many Casement stock-show victories in carload cattle competition and was present on such great occasions as when Dan Casement's colorful portrait was unveiled among those of other livestock industry immortals in the gallery of the prestigious Saddle and Sirloin Club in Chicago.

He is the author of the standard breed history, The Hereford in America, which has gone through three editions and is on reference shelves in every country in which Herefords are in significant production. He also has written portions of several other books, and articles by him have been published in several other countries.

J. EVETTS HALEY is at once a cattle rancher in the old tradition, who could ride and rope with the best, and a premier historian of the range country. Although he says that he now prowls the pastures at a slow pace and always on a gentle horse, his devotion and dedication to the cattle country and to the men who made it what it was continues undiminished. This led him to come to know and respect Dan Casement, whose instincts as an individual and philosophy as a tough-fibered, patriotic American exemplified attributes which he rated as highest of all.

Mr. Haley first earned national recognition as the author of The XIT Ranch of Texas, which was followed a few years later by Charles Goodnight, Cowman and Plainsman, considered by many to be the best range country biography ever written. Then came his George W. Littlefield, Texan; Jeff Milton: a Good Man With a Gun; and Fort Concho and the Texas Frontier. In addition to still other books, his vivid prose has brought to life scores of articles on the Southwest, which has always been his home.

R0l266 96659

R0127855581 txr   T
                     B
                     <3370
Ornduff, Donald R
Casement of Juniata : as a
    man and as a stockman